Finger Millet

Finger millets are gaining popular recognition as a superior source of phytochemicals, micronutrients such as calcium, phosphorous, and iron, and dietary fibre, when compared with cereals such as wheat and rice. *Finger Millet: Nutritional Properties, Processing, and Utilization* provides comprehensive information about the grain, including details of its different components such as starch, protein, bioactive metabolites, and their use in food products. It also discusses the effects of processing on their nutritional profile, as well as its utility as a gluten-free wheat substitute. This book will be a comprehensive reference tool for researchers, students, nutritionists, food technologists, policymakers, and agricultural practitioners.

Cereals: Science and Processing Technology
Series Editors: Sneh Punia and Manoj Kumar

Maize
Nutritional Composition, Processing, and Industrial Uses
Edited by Sukhvinder Singh Purewal, Pinderpal Kaur, Sneh Punia, Kawaljit Singh Sandhu, Surender Kumar Singh, and Maninder Kaur

Wheat Science
Nutritional and Anti-Nutritional Properties, Processing, Storage, Bioactivity, and Product Development
Edited by Om Prakash Gupta, Sunil Kumar, Anamika Pandey, Mohd. Kamran Khan, Sanjay Kumar Singh, and Gyanendra Pratap Singh

Oat (Avena sativa)
Production to Plate
Edited by Maharishi Tomar and Prabha Singh

Finger Millet
Nutritional Properties, Processing, and Utilization
Edited by Priyanka Kajla and Vandana Chaudhary

For more information about this series, please visit: www.routledge.com/Cereals/book-series/CSPT

Finger Millet
Nutritional Properties, Processing, and Utilization

Edited by
Priyanka Kajla and Vandana Chaudhary

CRC Press
Taylor & Francis Group
Boca Raton London New York

CRC Press is an imprint of the
Taylor & Francis Group, an **informa** business

Designed cover image: Shutterstock

First edition published 2026
by CRC Press
2385 NW Executive Center Drive, Suite 320, Boca Raton FL 33431

and by CRC Press
4 Park Square, Milton Park, Abingdon, Oxon, OX14 4RN

CRC Press is an imprint of Taylor & Francis Group, LLC

ISBN: 978-1-032-87874-4 (hbk)
ISBN: 978-1-032-87875-1 (pbk)
ISBN: 978-1-003-53511-9 (ebk)

DOI: 10.1201/9781003535119

Typeset in Palatino
by KnowledgeWorks Global Ltd.

Contents

Preface

Finger millet (*Eleusine coracana*), commonly known as Ragi, has been an integral part of traditional agriculture and is utilized as both food and feed in different parts of the world. In the last decade, millets have gained recognition due to their excellent nutritional profile, which enables them to thrive in harsh climatic conditions and thus play an important role in combating malnutrition and promoting environmental sustainability. This book is born out of the growing trend of millet utilization, more than just as a feed or an underutilized crop. It is a crop of the future as it is a hub of quality nutrients, and its climate resilience attributes make it a potential food crop in ensuring food security.

This book, titled *Finger Millet: Nutritional Properties, Processing, and Utilization*, is part of the Cereals: Science and Processing Technology Series and is edited by Dr. Priyanka Kajla and Dr. Vandana Chaudhary, under the series editorship of Dr. Sneh Punia Bangar and Dr. Manoj Kumar. The purpose of this book is to provide a comprehensive overview of the different aspects of finger millet and its utilization. The book is divided into 11 chapters, covering important information about updated research and applications. Chapter 1 includes a comprehensive overview of finger millet, encompassing its botanical characteristics, global and regional production trends, socio-economic importance, and the role of government policies to bring this crop into the mainstream agricultural food production system. Chapter 2 includes a detailed nutritional profile of finger millet, and the comparative nutritional significance of other millets is discussed at length. Chapter 3 delves into the non-nutritive or antinutritional factors present in finger millet. Chapter 4 explores bioactive compounds present in finger millet, focusing on polyphenols, antioxidants, and so forth, and their health benefits. Chapter 5 discusses the traditional and modern processing techniques of finger millet, such as milling, malting, fermentation, and extrusion. The main aim of these processing technologies is to reduce antinutritional factors, improve digestibility, and augment their utilization in the food processing sector. Chapter 6 outlines the importance of finger millet starch. It includes modification of starch to improve its functional properties, such as water absorption, emulsification, and heat stability, to enhance its utilization in different food systems. Chapter 7 outlines the impact of different processing technologies on the nutritional quality of finger millet. Chapter 8 reviews the role of bioactive compounds as therapeutic agents and their role in combating several lifestyle disorders such as diabetes, stress, and cardiovascular diseases. Chapter 9 focuses on the usage of finger millet in prebiotic and probiotic foods and the genetic modification of finger millet to enhance its food applications. Utilization of finger millet as a prebiotic or probiotic food enhances both nutrition and gut health, offering numerous health benefits along with genetic modification, and holds the potential to enhance the food application, improving the nutritional quality and addressing global food security. Chapter 10 outlines the development of different types of value-added products from finger millet, including bakery foods, weaning foods, health supplements, snacks, and beverages, catering to the growing consumer demand for gluten-free, nutrient-dense, and functional foods. Chapter 11 includes the latest trends or advancements in the processing of finger millet. It explores the latest technologies used to process various products with excellent nutritional and functional properties.

Our objective is to provide a comprehensive reference for researchers, students, nutritionists, food technologists, policymakers, and agricultural practitioners. We extend our sincere thanks to the contributors for their valuable work and commitment to advancing knowledge in this field. Each chapter is authored by experts in their respective fields, ensuring a rich and diverse representation of knowledge.

Contributors

Krishna Aayush
Department of Food Science and Technology, Graphic Era (Deemed to be University)
Dehradun, India

Reza Abedi-Firoozjah
Department of Nutrition, Faculty of Health and Nutrition Sciences, Yasuj University of Medical Science
Yasuj, Iran

Rahul Agarwal
School of Biosciences and Bioengineering, D.Y. Patil International University
Akrudi, India

Manisha Anand
Department of Food, Agriculture and Bioresources, Asian Institute of Technology
Bangkok, Thailand

Behnam Bahramian
Department of Food Science and Technology, Faculty of Nutrition and Food Science, Tabriz University of Medical Sciences
Tabriz, Iran

Vandana Chaudhary
Department of Dairy Technology, College of Dairy Science and Technology, Lala Lajpat Rai University of Veterinary and Animal Sciences
Hisar, India

Prafull Chavan
School of Biosciences and Bioengineering, D.Y. Patil International University
Akrudi, India

Navnidhi Chhikara
Department of Food Technology, Guru Jambheshwar University of Science & Technology
Hisar, India

Sohani Dadwal
Department of Biotechnology Engineering and Food Technology, University Institute of Engineering (UIE), Chandigarh University
Gharuan, Mohali, India

Priya Gupta
Department of Genetics and Plant Breeding, Institute of Agricultural Sciences, Banaras Hindu University
Varanasi, India

Udipta Hazarika
Department of Food, Agriculture and Bioresources, Asian Institute of Technology
Bangkok, Thailand

Indu
Master of Nutrition and Population Health, Deakin University
Melbourne, Australia

Shiksha Jangra
Department of Food Technology, Guru Jambheshwar University of Science & Technology
Hisar, India

Priyanka Kajla
Department of Food Technology, Guru Jambheshwar University of Science & Technology
Hisar, India

Sanjaya Karki
Department of Food, Agriculture and Bioresource, Asian Institute of Technology
Bangkok, Thailand

Simmi Ranjan Kumar
Department of Biotechnology, Faculty of Science, Mahidol University
Bangkok, Thailand

Palak Mahajan
Department of Biotechnology Engineering and Food Technology, University Institute of Engineering
Chandigarh University, Mohali
Punjab, India

Angelika Nelson
School of Applied and Life Sciences, Uttaranchal University
Dehradun, India

Anil Panghal
Department of Food Processing & Engineering, Chaudhary Charan Singh Haryana Agricultural University
Hisar, India

Amrita Poonia
Department of Dairy Science and Food Technology, Institute of Agricultural Sciences, Banaras Hindu University
Varanasi, India

Prince
Dairy Chemistry Division, ICAR-National Dairy Research Institute
Karnal, India

Himani Priya
Division of Microbiology, ICAR-Indian Agricultural Research Institute
Hazaribagh, India

Rinku
Department of Dairy Technology, College of Dairy Science and Technology, Lala Lajpat Rai University of Veterinary and Animal Sciences
Hisar, India

Rahul Saxena
Department of Food Technology, School of Engineering & Technology, Jaipur National University
Jaipur, India

Shiv
Department of Dairy Technology, College of Dairy Science and Technology, Lala Lajpat Rai University of Veterinary and Animal Sciences
Hisar, India

Gurvendra Pal Singh
Department of Food Technology and Nutrition, Lovely Professional University
Phagwara, India

Pawan Kumar Singh
Department of Genetics and Plant Breeding, Institute of Agricultural Sciences, Banaras Hindu University
Varanasi, India

Ranjit Singh
Agricultural Structures & Process Engineering (AS&PE), ICAR-Indian Agricultural Research Institute
Jharkhand, India
and
Department of Food, Agriculture and Bioresource, Asian Institute of Technology
Bangkok, Thailand

Ronak Tanwar
Department of Food Processing & Engineering, Chaudhary Charan Singh Haryana Agricultural University
Hisar, India

Milad Tavassoli
Department of Nutrition, Faculty of Health and Nutrition Sciences, Yasuj University of Medical Science
Yasuj, Iran

Amit Verma
M. Tech. Scholar, Department of Dairy Science and Food Technology, Banaras Hindu University
Varanasi, India

1

Finger Millet: General Aspects

Rinku, Amit Verma, Vandana Chaudhary, and Priyanka Kajla

1.1 Introduction

Millets are a diverse collection of compact grain grasses and are widely recognized as members of the grass family, belonging to the *Poaceae* family. They are cultivated as staple grains for livestock feed and human consumption. Asia has a long history of millet farming, dating back thousands of years, as evidenced by ancient documents. These traditional grains are nutrient dense and have become trendy due to their nutritional advantages, green practices, and culinary applications. Various millet varieties such as pearl millet, sorghum, foxtail millet, little millet, barnyard millet, kodo millet, proso millet, and finger millet are known for their unique nutritional characteristics and cooking applications. Looking at the balanced diet angle, millets are the suitable choice due to their high dietary fibre, minerals, antioxidants, and gluten-free nature. Majorly, finger millet, little millet, barnyard millet, and sorghum are cultivated in India. These varieties of millets are grown predominantly in drought-prone areas, including the arid and semi-arid landscapes of Rajasthan, Tamil Nadu, Gujarat, Maharashtra, and Andhra Pradesh. Specifically, in Asian and African regions, millets dished out as a staple food (Porwal et al., 2023).

Brief cultivation period, pest proofing, disease immunity, and sustainable output in arid climates make millets important crops in drought-prone and lush regions. Millets are neglected crops and have vast possibilities to transform millet grains into nutrient-enriched food and beverages in emerging economies (Devi et al., 2014). Additionally, millets contribute to eco-friendly cultivation practices and ecological protection. With fewer chemical fertilizer input, millets yield well and help in reducing the negative impact of agriculture on environment and eventually the soil health will improve (Porwal et al., 2023). In the recent years, the popularity of millets increased globally. The International Year of Millets 2023 recognizes the vital role of millets in achieving sustainable development objective (Theuri & Burkhart, 2023).

Governments, NGOs, and research institutions are collaborating to promote millet farming communities by raising awareness, developing

DOI: 10.1201/9781003535119-1

improved varieties, and implementing supportive policies. Encouraging millet consumption can help address nutrition challenges in India (Porwal et al., 2023).

Millets are generally classified into two categories, major millets and minor millets. The terms "major millets" and "minor millets" classify millet varieties based on their agricultural significance, dietary consumption, and economic impact. Sorghum, finger millet, and pearl millet are considered as major millets, while minor millets include little millet, kodo millet, barnyard millet, foxtail millet, and proso millet (https://chandigarhenvis.gov.in/sites/default/files/pdffiles/NL10122022.pdf).

Finger millet (*Eleusine coracana* L.) is a significant grain widely cultivated across India and Africa and consumed as a staple food by a large population in these countries. Millet is among the top six cereal crops produced in India. Millet seed coat is brick red in colour and generally consumed whole for cooking muddle, *Ambali,* and roti (Devi et al., 2014). Finger millet serves as a dietary staple food across various regions of African and South Asian countries. It is thought to have been cultivated around 4000 years ago. It is also considered an emergency food crop as it is conveniently kept for periods of shortage (FAO, 2016). Millet grains are highly digestible, nutritive, and adaptive, and cooking characteristics are similar to rice, milled to prepare porridge or flour or utilized in baking. Apart from this, finger millet is also utilized for making booze and beer. Considering the nutrient aspect, finger millet is a rich provider of essential nutrients and vital micronutrients. Because it is free from gluten and has a low glycaemic index (GI), it is advisable for individuals suffering from diabetes and celiac disease (Tripathi & Vyas, 2023).

From an Ayurvedic standpoint, finger millet helps in weight management, plant-based option for vegans, lowers triglyceride levels, reduces hyper tension, slows down skin aging, benefits pregnant women, promotes healthy child growth and weaning, reduces the risk of gallstones, and helps combat diseases or reduces the risk of cancer (Chandra et al., 2016).

1.2 Origin and Domestication of Finger Millet

Finger millet is thought to have arisen in Africa, boasts a long history before reaching the subcontinent of India. Its widespread crop production across the area of Africa makes it a vital staple for many small farmers, ensuring food security and sustainability. Millet is also cultivated in rural areas of China, India, and Nepal. Millet cultivation in the Korean Peninsula dates back approximately 3500–2000 BC. Yajurveda, an ancient religious text, signs the cultivation and utilization of millets in Asian region, tracing back to 4500 BC. Finger millet is originated from selective breeding of *E. coracana* subsp. *Africana* (Yadav et al., 2024).

Finger millet is a cultivated grain native to Africa, and in prehistoric times, widespread in Asia. Finger millet is also known as ragi in India and dagusa in Ethiopian language. While its wild ancestor, *Eleusine africana* (Kennedy-O'Byrne), is well documented, the precise African region where cultivation first began remains uncertain. From a habitation standpoint, finger millet is typically cultivated in hilly regions above 900 m. Apart from hilly region cultivation, in Asia, especially in India, millets are cultivated in plain regions.

Finger millet is less resilient than other millets, growing in nutrient-rich and humid soil, and still has exceptional storage qualities. However, finger millet grains are pest and spoil resistant and have a lifespan of up to 10 years. Generally, finger millet is cultivated in areas with 800–1000 mm rainfall but also survive on 500 mm rainfall. Finger millet is cultivated as an initial crop as it doesn't grow well in depleted land.

The oldest recorded historical evidence of finger millet cultivation dates back to around 2000 BCE. During the 6000 BCE, in Khartoum area, evidence of sorghum cultivation was found and the earliest verified specimen found in India around 2000 BC through commerce pathways from Yemen.

Finger millet is the fourth major cereal crop cultivated in Nepal due to its high margin and low external input compared to other agricultural crops with a cultivation area of 274,350 ha and an annual output of 302,000 metric tonnes. Finger millet grains are also utilized to brew ethnic alcoholic beverages due to its cultural importance in certain ethnic communities; it holds the potential to enhance food availability, support income, help in livestock production, and contribute to the holistic development of small farmers.

In various religious fasts, finger millet porridge is eaten due to its purity and sattvic qualities. The food choices during fasts reflect a strong connection between food habits and faith. Sattu, a traditional dish prepared by blending millet and barley flour, is offered as a holy tribute to Lord Shiva and Lord Vishnu on the occasion of Akshaya Tritiya with fruit juices. Finger millet is also an integral part of various culinary dishes, including Ven Pongal. Millet-based dishes are deeply connected to traditions, especially during the four-day festival of Pongal in January, dedicated to Lord Indra and Lord Bhaskar. On the final day, millet-based food is served due to the cultural significance of this grain (Yadav et al., 2024).

Finger millet grains are processed along with testa, which is a rich source of dietary fibre and essential minerals, to produce flour. The flour is utilized for making unleavened breads, thin porridge, and dumpling. Regular consumption of finger millet helps in lowering the risk of cardiovascular disease, insulin resistance diabetes, and abdominal tumours (Chandra et al., 2016). Tribal communities allocate three-quarters of their millet yield for daily consumption as part of their tradition. Around two-thirds of people have yearlong availability of millets in the household but efforts are still needed to ensure year along availability to remaining one-third of people. In most household, finger millet is consumed as a source of high energy food, but they are unaware of the role of millets in controlling blood sugar levels.

1.3 Botanical Description and Agronomic Features

Finger millet is classified under the *Gramineae* (or *Poaceae*) family. The botanical name for finger millet is *Eleusine coracana* which belongs to the genus *Eleusine* consists of about nine species (Fig. 1.1). Among them, *E. coracana* is widely cultivated and utilized in human diets (Table 1.1).

This categorization is important as it includes finger millet as a part of the *Poaceae* family, which is one of the most fundamental groups for cereal grasses, including key staple grains such as maize and rice (Dangwal et al., 2024). Finger millet and its variants are classified within the *Chloridoideae* subfamily. The crop is grouped into the genus *Eleusine*, comprising eight species that encompass both short-lived and long-lasting varieties. The height of the finger millet plant ranges from 40 to 150 cm and needs five to six months to reach full maturity. The flowering structure of finger millet consists of multiple spikes, varying from 3 to 20, arranged in a claw-like pattern. It appears similar to fingers, eventually named it "finger millet". Each stalk contains around 70 alternative tiny florets along the central axis, and each floret produces four to seven seeds. The diameter of seeds ranges from 1 to 2 mm. The colour of the seed coat varies significantly from variety to

FIGURE 1.1
Finger millet.

TABLE 1.1
Botanical Classification of Finger Millet

Botanical Classification	
Kingdom	Plantae
Division	Magnoliophyta
Class	Liliopsida
Order	Poales
Family	*Poaceae*
Genus	*Eleusine*
Species	*E. coracana*

variety, such as orange, purple, red, brown, and white. The floral structure of the finger millet shows wide variation and may be due to farmers' selective breeding. Based on the floral arrangement, finger millet can be classified into five races. The coracana race, which closely resembles the *africana* subspecies, features well-developed central floral stems. The vulgaris race displays a bloom pattern. The Cockscomb finger millet, referred as compacta race, has curved spikes with separated lower branches. The large spikelets were characterized in plana race, and the elongata race develops long spider spikes that curved inwards upon reaching maturity, with lengths extending up to 24 cm (Dida & Devos, 2006).

Finger millet is a low-latitude and warm climate cereal which can resist extreme heat and dryness which suggests its capability to grow in the environment where other cereals such as maize, rice, and wheat are difficult to grow. Finger millet is used as an economically accessible food crop due to its long-lasting resilience as it could be stored safely for prolonged periods and free from pest attacks (Rathore et al., 2019). Finger millet can withstand different farming climates when cultivated as a low water-demanding crop in places with scarce water supply. After foxtail millet, pearl millet, and sorghum, finger millet is the fourth most widely cultivated crop in low rainfall areas. It may offer maximum productivity with minimal resources because it is a warm climate dryland crop that can grow on low fertility soil, in elevated regions, and is resistant to drought and soil salinity. Insect infestations, plant diseases, and the harsh climate do not have significant effect on finger millet. While June or January is the period for collection, February and August are the sowing seasons (Sahoo et al., 2024).

Finger millet is versatile to diverse weather patterns, which is the explanation for its broad global presence. It can be cultivated in altitudes ranging from sea level up to 2100 m. It grows well in the regions where the temperature is between 20°C and 30°C. Moderate rain of 500–700 mm per annum is capable to support its cultivation, making it ideal for drought-prone regions (Babypriyanka et al., 2025).

When it comes to soil, finger millet performs well in properly aerated, fertile to light-textured soil. It tolerates low pH soils with a pH range of 4 to 8, which makes it resilient to a wide variety of soil types, including those regarded as poor quality and inadequate for conventional farming. Its extensive rooting characteristic allows it to extract nutrients and moisture from deeper soil layers, providing it with a unique ability to withstand a harsh environment (Maqbool et al., 2022). The adaptability of finger millet in both climate and soil conditions presents it as a prime selection for eco-friendly farming. Its ability to grow in challenging environments helps in ensuring a reliable food supply, particularly the areas that are increasingly susceptible to the influence of a changing climate (Babypriyanka et al., 2025).

1.4 Global and Regional Production Trends

According to Food and Agriculture Organization Corporate Statistical Database 2023, the world millet production in 2023 was 315.96 lakh metric tonnes, and in India, it was 135.07 lakh metric tonnes (Fig. 1.2). India produces around 38.4% of the world's millet production (FAO, 2023a). USD 34.32 million worth of millet products and USD 26.97 million worth of millets were exported during 2021–2022 (Press Information Bureau, 2022, RU-16-01-0158-261222). In FY24, millet production in India was 15.38 million metric tonnes, and 12.19 million hectares were under cultivation (APEDA, 2024).

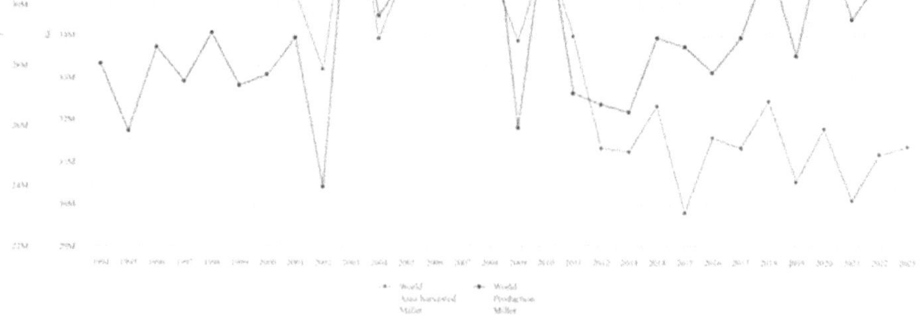

FIGURE 1.2
Production and yield of millets in the world.

TABLE 1.2

Country-Wise Finger Millet Production Statistics, Area under Production, and Yield and Relative Importance

Countries	Relative Importance (%)	Total Production Area (ha)	Total Production (tons)	Yield (t/ha)
India	1.3	1,004,460	1,755,060	1.75
Ethiopia	100	48,051	1,218,582	2.54
Nepal	100	263,261	314,225	1.19
Uganda	94	141,600	197,091	1.39
Tanzania	30	81,000	97,500	1.2
Kenya	45	53,285	68,850	1.29
Zimbabwe	30	43,349	13,023	.3
Zambia	60	39,454	27,003	.68
Malawi	60	34,423	29,359	.85
Eritrea	20	15,000	5000	.33
Mozambique	20	10,695	4420	.41
Rwanda	100	10,637	5067	.48
Finger millet	**Total/mean**	**2,177,675**	**3,735,180**	**1.03**

Finger millet originates from the East African highlands and extensively farmed in more than 25 countries including East China, Tanzania, Nepal, India, China, and Kenya. It is believed that Uganda or surrounding locations serve as the point of origin of *E. coracana*, and it was introduced to India approximately 3000 years ago. Karnataka dominates the finger millet industry in India, contributing 58% global sharing, but only certain groups are aware about its wellness benefits (Rathore et al., 2019). In Asia and Africa, when it comes to cereals, finger millet holds the fourth position. Annually, the total finger millet overall yield is around 4.5 million tonnes, with an overall output of 1.67 million tonnes and a typical productivity of 1747 kg/ha, finger millet is cultivated over 1.01 million hectares in India (Aman et al., 2023). As comprehended in Table 1.2, the total area under finger millet cultivation is approximately 2.17 million hectares, resulting in a total production of 3.73 million tonnes.

In the Southern states of India, finger millet is grown widely due to favourable environmental conditions (Kumar et al., 2024). Within India, Karnataka dominates, accounting for 56.21% of the cultivated area and 59.52% of the total output. It is followed by Tamil Nadu, contributing 9.94% to the area and 18.27% to the production; Uttarakhand with 9.40% area share and 7.76% production; and Maharashtra, covering 10.56% of the cultivation area and producing 7.16% of the total yield (Rajeshwari et al., 2025). As per the data obtained from APEDA (Table 1.3), total area under millet cultivation in

TABLE 1.3

Area and Export Data of Millets in India

Parameter	Millets
Production in FY24 (mil MT)	15.38
Area under cultivation in FY24 (mil ha)	12.19
Volume of export FY24 (lakh MT)	1.46
Exported in FY24 (USD mil)	70.89

Source: APEDA (2024).

financial year 2024 was 12.19 million hectares, yielding 15.38 million metric tonnes of production. India exported 1.46 lakh metric tonnes of millets in FY24.

1.5 Traditional Uses and Cultural Importance

Finger millet has been utilized as a staple food, especially in areas such as India and Nepal, in which it is utilized to make wholesome food, including unleavened bread, cereal mush, and cultured drinks. Apart from nutritional significance, finger millet is esteemed for its outstanding dietary composition, high in calcium, iron, and dietary fibre, contributing to growth in children, prenatal care, and bone density. Traditionally, it is involved in ceremonies and celebrations, frequently included in spiritual gifts and ancestral rites. Moreover, its dry climate-tolerant characteristic makes it an essential cereal for nutritional stability in dry and moderately arid areas, where its farming supports the livelihood of local cultivators and regional markets. Therefore, finger millet is more than a nutritional provider but a backbone of tradition, wellness, and strength in many societies.

Finger millet Huri Hittu, a dish cooked with puffed finger millet, is high in minerals and micronutrients. In South India, finger millet has been sprouted for culinary applications in all historical records. It has outstanding brewing properties, and the fermented millet has an acceptable taste and superb aroma, and durable freshness (Ankita & Seth, 2025). Finger millet *Mudde* is the traditional cuisine eaten by the locals in Karnataka for the first meal. Finger millet is an essential harvest in Karnataka, was cooked by homemakers utilizing kodo millet, foxtail millet, and little millet. These meals were tasty and offered with sambhar, which was prepared from a range of regional legumes, with hyacinth beans excelling for strong flavourful aroma and dietary benefits. Housemakers in remote place believed that this soup and millets combination is highly healthy and appetizing (Hazareesingh, 2021).

Ambali is a fermented millet porridge, South Indian dish prepared from aged millets. It is prepared by blending finger millet flour with water to create a dense dough, which is then heated and cultured. *Koozh* is a cultured millet-infused dish indigenous to Tamil Nadu region, well recognized for its unique taste and health benefit and frequently consumed as a first meal. Finger millet is also used for roti making. Highly recognized millet-based meal prepared by the Monpa community of Arunachal Pradesh is known as *Zan* and is prepared with finger millet and green vegetables (Ankita & Seth, 2025). *Landa* and *Pej* are the regularly enjoyed cultured finger millet beverage considered as essential drink by the Gadaba community of Odisha. *Landa* includes sprouting and popping, and particularly consumed in the nighttime for energy boost. It has proven a highly nutritious diet via sprouting and bursting process, which enhances the accessibility of essential nutrients such as amino acid and carbohydrates. Prepared using finger millet flour, *Pej* is consumed for midday meal and evening meal. It is a popular drink that is necessary to be offered during noon feast and dinner (Mishra et al., 2021).

In specific spiritual fasting and traditions, finger millet-infused meals, like Kodo ko Jhol, are eaten because they are viewed as wholesome and sattvic. This tradition shows the profound bond between food-related choices and faith. Finger millets have performed a significant role in societal engagement in various traditions. The Kuthiyottam festival, celebrated in the region of Kerala, is an example of crop festival related to millets. In November and December, a gathering, paying tribute to Lord Vishnu, is performed. Cultural performance known as "Kuthiyottam" is presented during the celebration, and it includes detailed rituals and millets dedicated to the supreme entity. The Yimchunger community of Nagaland celebrates Metumniu and Tsungkamniu in honour of millet harvesting completion (Jeyaram et al., 2010). Makar Sankranti, primarily celebrated in North Indian states of India, emphasizes the connection between crop collection and cultivation-based growth and plentiful supply. Finger millet is commonly given as sacred food in the festival of Gaddi. Moreover, the finger millet provides positive strength in temple and dense mixture prepared using millet flour is applied to heal heat wounds (Ragupathy et al., 2016). Throughout marriages, it is common for various communities in the region of Madhya Pradesh to apply a blend of turmeric and finger millet flour on the groom and bride (Rana, 2022).

Finger millet application as ancient remedies is age old as its farming. It has been commonly used in Ayurveda, in India and in African medicinal systems. In Ayurveda, finger millet is commonly advised to individuals with high blood sugar levels because of its low GI, high dietary fibre, and complex carbohydrate. The gradual absorption of glucose helps improve diabetes regulation. This attribute ensures that it is the ideal food for people suffering from diabetes and reduces the risk of Type 2 diabetes. The high heart-friendly fat and richness of magnesium and potassium in finger millet help in reducing the risk of heart problems. Minerals present in finger millet help in controlling blood circulation and enhancing cardiovascular performance.

Moreover, the plentiful amounts of dietary fibre aids in lowering the level of low-density lipoprotein, lowering the risk of deposits in the blood vessels.

The protein presents in finger millet is regarded as high quality as it contains essential peptides such as valine, lysine, and threonine. Finger millet is prescribed for health issues related to hypertension, liver diseases, respiratory conditions, and cardiac fragility. Additionally, breastfeeding mothers who lack milk production are recommended to add millets to their diet. In case one consistently eats finger millet, it has the potential to prevent undernourishment, chronic diseases, and keep early maturing under control (Chand et al., 2024). The health benefits of finger millets and its products are highlighted with its emerging role in the pharmaceutical industry. The acetylated finger millet (AcFM) and oxidized finger millet (OxFM) forms of starch extracted from finger millet are used as fillers in capsules and are widely used as a drug carrier. AcFM and OxFM show increased tapped and bulk density, good flowability, and solubility (Chand et al., 2024).

1.6 Government Initiatives and Promotion Efforts

The year 2023 was designated as the International Year of Millets by United Nations General Assembly. The suggestion was first submitted by India and later, it was backed by nearly 72 nations. This global recognition was a result of decade-long policy evolution within India aiming at getting millets back into mainstream national and international food systems. Now, Indian administration is observing International Year of Millets by engaging individuals of India in millet-based product and premium processed foods. Atmanirbhar Bharat Abhiyan, an initiative by the Government of India, sanctioned a national programme called "Production Linked Incentive Scheme for Food Processing Industry" on 31 March, 2021. The primary goal of this programme encompasses assistance for establishment of worldwide agro-processing industry and backing local enterprise of food products in the global markets. Particular culinary categories having strong expansion prospects have been recognized for offering assistance under the programme. Millets were included in POSHAN Mission Abhiyan, a national campaign aimed at nutritional security. Distribution of this information can be communicated to people by the administration through public relation office. Indian Council of Agricultural Research–Indian Institute of Millets Research (ICAR-IIMR) has formulated a number of millet-based food items like millet snacks, finger millet cakes, and high-quality sorghum almond biscuits. These newly formulated millet-based food items were showcased at national seminars and were presented for sensory evaluation to the experts and policymakers (Chand et al., 2024). Dissemination of information regarding the above-mentioned efforts has resulted in public awareness and introduction of millets in daily

TABLE 1.4
Policy Initiative Related to Millet's Promotion in India

Year	Intervention
2012	Initiative for Nutritional Security through Intensive Millet Promotion (INSIMP)
2013	National Food Security (NFS) Act covers "coarse grains"
2017	NITI Aayog of Government of India releases the National Nutrition Strategy (NNS) for "Nourishing India" and recommends that the Ministry of Agriculture & Farmers' Welfare strengthens cereal productivity and production diversity – including the production of "coarse" cereals such as millets
2018	1. Millets officially declared as "Nutri-cereals" 2. Millets made part of the National Food Security Mission (NFSM) 3. Government of India declared 2018 as the "National Year of Millets" 4. Government of India sent a proposal to United Nations for declaring 2023 as the "International Year of Millets" to promote greater production and consumption of millets
2021	United Nations General Assembly has approved and declared 2023 to be observed as the "International Year of Millets" (IYM)
2021	Government of India revises its guidelines for procurement, allocation, distribution, and disposal of coarse grains: Nine months for jowar and bajra, ten months for finger millet, and six months for maize. This would increase procurement and consumption of these commodities as the state would have more time to distribute these commodities in Targeted Public Distribution System/ Other Welfare Schemes Provision of inter-state transportation of surplus coarse grains through Food Corporation of India (FCI) is incorporated to cater for advance demand placed by consuming state before the start of procurement

Source: NITI Aayog (2023). www.niti.gov.in

foods. Table 1.4 summarizes the key policy milestones, demonstrating consistent efforts towards reintroducing millets like finger millet into mainstream agriculture activities, food security, and nutrition agendas.

1.7 Future Perspectives

Owing to climate change, water scarcity, increasing population, and increasing demand for sustainable agricultural practices, global agricultural landscape is confronting critical challenges. Millets, in particular finger millet, have emerged as a crop of strategic importance. Once a main dietary ingredient for communities like Mudra, the finger millet is underutilized. One of the primary reasons cited by the society for reduced farming of finger millet was change to farming of hybrid rice, as this crop has higher yield linked to greater financial benefits (Downs et al., 2022). Despite its rich nutritional profile indicated by high amount of calcium, dietary fibre, and thiamine, it is rarely eaten due to limited farming yield, even though it tastes appealing. For revitalizing, the finger millet in main diet requires a multifaceted approach exploiting the traditional wisdom with modern agricultural practices. Supply network of millets can be improved through proper distribution based on availability and demand factors. The cultivation and usage of millets must be improved with suitable regulatory actions as an extension of government programmes like the National Food Security Mission (NFSM) and Initiative for Nutritional Security through Intensive Millets Promotion (INSIMP). Collaborative studies may be carried out to confirm high yield methods for practical agriculture settings. Efficient utilization of feature focused germplasm available in seed reserves for breeding improvement and adaptive methods for improved crop yield along with government, corporate, and non-profit collaborations, and the setup of farming clusters may be encouraged (Shanmuka & Shireesha, 2023).

Emerging treatments and formulation techniques are required to improve the accessibility of trace minerals and elevate the value of finger millet-based food. Expansion of market for finger millet-based food product that can be consumed by consumers affected by gluten intolerance and blood sugar disorder is essential. Considering the ongoing difficulties concerning eco-friendly food production, atmospheric variations, and water shortage, combined with population surplus, curiosity has been emerged about millets. This situation has offered chance for cultivators, dietitians, and cuisine and fodder producers to involve themselves in study to grasp the dietary and functional identification of millet grains (Chand et al., 2024). The Indian administration has acknowledged the capability of millets and has implemented regulatory measures to encourage startups. This initiative involves promoting millets in public sector, offering monetary aid and grants to innovators, and developing supportive ecosystem for millet-based ventures to flourish. Business owners regularly work together with research institution and farming entities to create cutting-edge technologies and millet-based products. These partnerships support in enhancing the production methods, creating premium goods and boosting the standards and commercial viability of millet-based food (Shanmuka & Shireesha, 2023).

1.8 Conclusion

In conclusion, finger millet possesses notable relevance owing to its dietary and health benefit profile. It contains high levels of bone strengthening mineral, iron, and dietary fibre, rendering it a beneficial component for daily meal, particularly in the area where nutritional deficiency is common. Including finger millet within multiple products contributes variety in mean plans and aids in tackle dietary deficiencies, notably among kids, as well as expectant mothers (Das et al., 2024). Innovators are investigating creative methods to include millets into assorted cuisine offering to address the evolving buyer choices and nutritional security (Shanmuka & Shireesha, 2023). By improvement in its cultivation practices with national policies, finger millet emerges as a potent contributor in building a healthier food future.

References

https://chandigarhenvis.gov.in/sites/default/files/pdffiles/NL10122022.pdf

https://www.milletstats.com/apy-stats/

Aman, Z., Singh, S., Pandey, M., & Chaudhary, A. (2023). History of millets. *Millets: Forgotten Grains Regaining Prominence, 1.*

Ankita, & Seth, U. (2025). Millets in India: Exploring historical significance, cultural heritage and ethnic foods. *Journal of Ethnic Foods, 12*(1), 2.

Anonymous (2022). Paryavaran patra; Millets: Reviving Health and Heritage. https://chandigarhenvis.gov.in/sites/default/files/pdffiles/NL10122022.pdf

APEDA (2024). Indian millets. https://apeda.gov.in/IndianMillets

Babypriyanka, S., Hepziba, J. S., Pushpam, K. A., Pillai, A. M., Vijayalakshmi, R., & Theradimani, M. (2025). Exploring the bioactive components of millets for their nutraceutical potential. *Food Science and Biotechnology, 34*(3), 563–575.

Chand, S., Chauhan, S., Kumar, A., & Chauhan, M. (2024). Export potential of Indian millets status and prospects. *Millets,* 258.

Chandra, D., Chandra, S., & Sharma, A. (2016). Review of finger millet (*Eleusine coracana* (L.) Gaertn): A power house of health benefiting nutrients. *Food Science and Human Wellness, 5*(3), 149–155.

Dangwal, L. R., Rawat, M., & Lal, T. A (2024). Comprehensive Review on the nutrition, processing, phytochemical, and pharmacological properties *of Eleusine coracana* (L.) Gaertn (Poaceae). *Journal of Mountain Research, 19*(2), 43–52.

Das, P., Subba, R., Krishnan Kesavan, R., & Nayak, P. K. (2024). Finger millet: Processing innovations and value addition for enhanced utilization and market potential. AgriTech Today, 2(3), 6–10.

Devi, P. B., Vijayabharathi, R., Sathyabama, S., Malleshi, N. G., & Priyadarisini, V. B. (2014). Health benefits of finger millet (*Eleusine coracana* L.) polyphenols and dietary fiber: A review. *Journal of Food Science and Technology, 51,* 1021–1040.

Dida, M. M., & Devos, K. M. (2006). Finger millet. In *Cereals and millets* (pp. 333–343). Springer.

Downs, S. M., Kapoor, R., Merchant, E. V., Sullivan, T., Singh, G., Fanzo, J., & Ghosh-Jerath, S. (2022). Leveraging nutrient-rich traditional foods to improve diets among indigenous populations in India: Value chain analysis of finger millet and kionaar leaves. *Foods, 11*(23), 3774.

Food and Agriculture Organization (FAO). (2016). *The state of food insecurity in the world 2015. Meeting the 2015 international hunger targets: Taking stock of uneven progress.* Food and Agriculture Organization Publications, Rome.

Food and Agriculture Organization (FAO). (2023a). Crop and livestock products. https://www.fao.org/faostat/en/#data/QCL/visualize

Food and Agriculture Organization (FAO). (2023b). Six reasons to bring millets to the market! https://www.fao.org/fao-stories/article/en/c/1628815/.

Hazareesingh, S. (2021). 'Our grandmother used to sing whilst weeding': Oral histories, millet food culture, and farming rituals among women smallholders in Ramanagara district, Karnataka. *Modern Asian Studies, 55*(3), 938–972.

Jeyaram, K., Romi, W., Singh, T. A., Devi, A. R., & Devi, S. S. (2010). Bacterial species associated with traditional starter cultures used for fermented bamboo shoot production in Manipur state of India. *International Journal of Food Microbiology, 143*(1–2), 1–8.

Kumar, B., Singh, M., Kumar, A., Kumar, D., Singh, A. K., Kumar, R., & Kumar, S. (2024). Finger millet scientific cultivation and its uses in India. In *Millets-Rediscover Ancient Grains*. IntechOpen.

Maqbool, S., Hassan, M. A., Xia, X., York, L. M., Rasheed, A., & He, Z. (2022). Root system architecture in cereals: Progress, challenges and perspective. *The Plant Journal, 110*(1), 23–42.

Mishra, A. K., Bajpai, R., & Swain, A. (2021). Finger millet-based staple beverages consumed by the Gadaba ethnic community of Odisha, India: Preparation and nutritive characteristics. *Biodiversitas Journal of Biological Diversity, 22*(5).

Porwal, A., Bhagwat, G., Sawarkar, J., Kamble, P., & Rode, M. (2023). An overview of millets-the nutri-cereals: Its nutritional profile, potential health benefits and sustainable cultivation approach. *International Journal of Science and Research Archive, 10*(01), 841–859.

Press Information Bureau (2022). Significant rise in MSP over last five years helps raise production of Millets; boost farmers' income in the country. https://static.pib.gov.in/WriteReadData/specificdocs/documents/2022/dec/doc20221226147401.pdf?trk=public_post_main-feed-card_feed-article-content

Ragupathy, S., Dhivya, S., Patel, K., Sritharan, A., Sambandan, K., Gartaula, H., Sathishkumar, R., Khadka, K., Nirmala, B. C., & Kumari, A. N. (2016). DNA record of some traditional small millet landraces in India and Nepal. *3 Biotech, 6*, 1–19.

Rajeshwari, S., Narayanan, A., Sridevi, V., Nadarajan, S., & Pushparaj, S. (2025). Effect of liquid nanourea application on yield, post-harvest plant and soil nutrient status in finger millet (*Eleusine coracana* L.) production. *Journal of Advances in Biology & Biotechnology, 28*(3), 474–480.

Rana, N. K. (2022). The perpetual quest for 'authenticity' in Indian cuisine: An answer through history and folklore. *Digest: A Journal of Foodways and Culture, 9*(1).

Rathore, T., Singh, R., Kamble, D. B., Upadhyay, A., & Thangalakshmi, S. (2019). Review on finger millet: Processing and value addition. *The Pharma Innovation Journal, 8*(4), 283–291.

Sahoo, S. S., Anushree, R., Agarkar, B., Kshirsagar, R., & Pawar, V. (2024). Finger millet – morphological, chemical and biological properties. In *Nutraceutical potential of millets*, (Eds. Gupta, E., Maurya, N. K. & Yadav, L.) (p. 195). P.K Publishers & Distributors, Delhi.

Shanmuka, A., & Shireesha, K. (2023). An overview on value chain development in millets. *Biological Forum – An International Journal, 11*, 5.

Theuri, S., & Burkhart, S. (2023). Calling on all SNEB members to engage in the "International Year of the Millets 2023". *Journal of Nutrition Education and Behavior, 55*(1), 1.

Tripathi, T., & Vyas, S. (2023). From ancient grains to modern solutions: A history of millets and their significance in agriculture and food security. *International Journal of Home Science*, *9*(2), 72–78.

Yadav, S. P. S., Khadka, A., Katel, S., & Karki, R. (2024). Finger millet: A holistic exploration of its cultural, historical, religious, indigenous knowledge, and culinary significance. In: *Millets Traditions, Science and Technology in Nepal*, (Eds. Joshi, B.K., Shrestha, R.K., Ghimire, K.H., HB KC and A GC), 14–29. NAGRC, CCDABC and FAO; Kathmandu, Nepal.

2

Nutritional Profile of Finger Millet and Comparative Nutritional Significance over Other Millets

Udipta Hazarika, Manisha Anand, and Simmi Ranjan Kumar

2.1 Introduction

Finger millet (*Eleusine coracana* L. Gaertn) is a nutritious but underutilized cereal crop from the grass family, commonly grown in dry regions of Africa and Asia. It is the fourth most important type of millet and is packed with nutrients like vitamins, amino acids, calcium, and iron (Rakkammal et al., 2023; Rasri, Kumar, and Khampratueng 2024). There are several types of finger millet, which can be classified based on grain color as brown, light brown, or white (Brhane et al., 2022). The white variety is primarily used for baking, while the light brown and brown types are commonly used for porridge preparation. In Southern Africa, the brown variety is also widely used for brewing traditional beer (Dhanushkodi et al. 2023). Finger millet is a richer source of dietary fiber compared to other millets and major crops. This is due to its unique kernel structure, which consists of three layers—an outer seed coat, a middle endosperm, and an innermost embryo—and is distinguished by a five-layered testa that sets it apart from millets like foxtail, kodo, pearl, and proso millet (Kaur et al. 2024).

Finger millet is incredibly nutritious, containing more carbohydrates (65–75%) and proteins (6–13%) than wheat and rice. It is also rich in balanced proteins, calcium (220–450 mg), iron (3–20 mg), essential amino acids, and B vitamins. Finger millet provides 300–350 kcal of energy and is rich in phytochemicals, polyphenols, flavonoids, and tannins, making it valuable for food and nutraceutical applications (Kaur et al. 2024; Rasri, Kumar, and Khampratueng 2024). Through processing, the nutritional value of finger millet can be enhanced. It is also gluten-free and easily digestible, making it a good option for celiac disease patients (Abioye et al. 2022). It can produce resistant starch, which is not digested in the small intestine but is broken

 DOI: 10.1201/9781003535119-2

down by gut bacteria, leading to production beneficial short-chain fatty acids butyrate, acetate, and propionate. Similar to dietary fiber, resistant starch promotes gut health, helps control blood sugar levels, and supports weight management, making finger millet a valuable ingredient for functional foods. Finger millet, known for its excellent nutrition and versatility, is becoming a great alternative to common grains like wheat, rice, and maize in many food products (K et al. 2024). This chapter aims to provide a comprehensive analysis of the nutritional benefits of finger millet compared to other staple crops. It is structured as follows: First, it explores finger millet's composition, including its nutritional and antinutritional properties. Next, an in-depth discussion covers the bioactive compounds, starch and protein digestibility, and various processing techniques used to enhance the nutritional qualities of finger millet.

2.2 Composition and Nutritional Significance of Finger Millets in Comparison to Other Millets

2.2.1 Major Nutrients/Constituents

Finger millet is an excellent source of various macro- and micronutrients. Depending on the various species, nutritional quality and percentages vary. Nutritional profile of millets can be compared to that of rice and wheat, which makes them a great source of nutrients. A comparative study which focused on the mineral composition of various food crops found that millet had the highest levels of minerals (when compared to soybean, wheat, corn) in the following order: K>Mg>Ca>Fe>Zn>Mn>Cu>Mo>Li (Anandharamakrishnan, Rawson, and Sunill 2022). The major nutrients found in finger millet include carbohydrates, proteins, dietary fiber, and various minerals. Moreover, phytochemicals such as phenolic compounds and other bioactives are also present in the millet kernel. The nutritional profile may vary among the various species depending upon the environment, genotype, grain, and a range of other factors. Overall, millets comprise of about 60–78% carbohydrates, 6.0–12.30% protein, 1.90–4.30% fat, and 1.30–2.70% ash (Table 2.1) (Mishra, Kumar, and Srivastava 2024)

Carbohydrates are the major constituent in the nutritional profile of finger millet. It comprises about 55.0–65.0% in the form of starch. Of this, free sugar constitutes about 1.04% while non-starchy polysaccharide content is about 65.50%. It was reported that the carbohydrate profile of few varieties constituted about 0.04–0.60% lignins, 1.40–1.80% cellulose, 6.20–7.20% pentosans, and 59.20–61.20% starch (Shobana et al. 2013). The endosperm layer of the grain constitutes the highest percentage of starch, and the starch granules vary in size according to their variety. Furthermore, the starch

TABLE 2.1

Nutritional Comparison of Finger Millet with Millets and Wheat

Millets	Carbohydrate (%)	Protein (%)	Fat (%)	Ash (%)	Total Dietary Fiber (%)
Finger millet	66.0–75.0	7.1–10.5	1.3–1.8	2.01–2.9	9.5–11.2
Foxtail millet	60.1–70.5	11.0–12.5	3.6–4.5	2.7–3.2	9.0–11.0
Kodo millet	66.0–70.0	7.2–13.4	1.7–3.5	1.7–4.2	6.1–7.5
Proso millet	64.5–75.55	10.5–13.7	1.54–3.78	2.3–4.0	7.4–10.2
Barnyard millet	60.3–65.0	7.2–10.5	2.0–6.3	1.4–4.0	10.0–12.6
Little millet	66.8–75.5	10.1–13.4	3.6–4.3	1.2–4.2	7.1–9.3
Wheat (whole)	60.2–65.7	10.0–11.5	1.4–3.0	1.2–3.84	9.04–11.4

Source: Mishra, Kumar, and Srivastava (2024), Kumar, Tripathi, et al. (2021), Anal et al. (2024), Shobana et al. (2013), Anandharamakrishnan, Rawson, and Sunill (2022), and Sharma et al. (2021).

granules are highly organized in the corneous endosperm, whereas these are loosely packed in the floury endosperm part of the seed. This can lead to formation of larger granules and make it more prone to enzymatic digestion (Dida and Devos 2006; Shobana et al. 2013). Finger millet starch is comprised of amylose and amylopectin, with amylose content being around 16%. The dietary fiber content that is present in finger millet (11.50%) is significantly higher when compared to polished rice, brown rice and other millets such as foxtail, kodo, or little millet (Shobana et al. 2013). The total dietary fiber, soluble dietary fiber, and insoluble dietary fiber was found to be around 12%, 2%, and 11%, respectively. This makes a significant impact on the health of an individual, as high dietary fiber content is related reduction in postprandial glucose levels as well as providing a hypoglycemic effect (Singh and Raghuvanshi 2012).

Protein is the second major component present in finger millet. However, changes in the protein content vary significantly within species and varieties, which can be due to a myriad of factors such as genetic variation, geographical location, soil condition, etc. Millets, as a species, are a good source of proteins, with protein content in the range of 10–15% (Yousaf et al. 2021b). The distribution of protein content within the millet kernels are in the following order: endosperm—80% > germ—16% > pericarp—3% (Anandharamakrishnan, Rawson, and Sunill 2022). This significant presence of protein makes it a suitable candidate for production of value-added food products. A study compared the protein content of 16 finger millet varieties, and it found to be in the ranges between 4.89% and 15.58% (Singh and Raghuvanshi 2012). In finger millet, both prolamins and glutelins are major fractions that form the protein profile, as compared to other millet varieties where prolamin is the major fraction (Mishra, Kumar, and

Srivastava 2024). The quality of protein also plays an essential role in its digestion and bioavailability, which depends on the percentages of essential amino acids. The chemical score of finger millet was found to be 52, in comparison to pearl millet (63) and sorghum (37) (Chamoli et al. 2018; Shobana et al. 2013). Major cereals are deficient in lysine; however, millets, particularly finger and kodo millets, may contain lysine in the range of 2.20–5.50 g lysine/100 g. This can be due to the presence of certain fractions (albumin, glutelin, globulin) that contribute to the high concentration (Sharma et al. 2021; Yousaf et al. 2021b). Finger millet also has a higher concentration of sulfur-containing amino acids (methionine and cystine), as compared to milled rice. When compared to other millets, the essential amino acid content of finger millet was fairly balanced with higher concentrations of lysine, threonine, and valine. However, presence of tannin can impact the protein digestibility, which can be improved by supplementation for improved nutritional status, and improved digestibility (Shobana et al. 2013).

Lipids are an essential nutrient and most millets have comparatively higher quantity of lipids when compared to wheat. A higher percentage of lipid in the grain can lead to higher oxidation rate, and as a result the grain can spoil at a faster rate during storage (Anal et al. 2024). A study by Sridhar and Lakshminarayana (1994) compared the lipid composition of foxtail, proso, and finger millet, where it was found that the total lipid content varied significantly among various cultivars, with the highest being in foxtail millet (11%), followed by proso millet (9%), and around finger millet (5.20%). Furthermore, out of the total lipids, 80% are triacylglycerols, 14% phospholipids, and glycolipids accounted for 6% of the total concentration in finger millet (Sridhar and Lakshminarayana 1994). Specifically, finger millet has a lipid concentration of 70–72%, which are mostly triglycerides with sterol traces. Additionally, glycolipids and phospholipids consist of around 10–12% and 5–6%, respectively. As is the case with most millets, finger millet too has a higher concentration of lipid in the pericarp and germ. As a result, during decortication, lipid amount is drastically reduced. Other millets such as foxtail, pearl, barnyard have a higher fat content when compared to finger millet, which can assist in enhancing the storage life of finger millet grains (Saleh et al. 2013; Shobana et al. 2013).

The significance of micronutrients in the form of minerals and vitamins is quite exceptional as all the millets are rich in various micronutrients, which include potassium, calcium, iron, magnesium, zinc, etc. as well as B-vitamins such as niacin, riboflavin, thiamine, etc. although it varies according to the various millet species. Finger millet, in particular, is a great source of calcium and has the highest concentration among all millets (364 mg/100 g) (Kumar, Tripathi et al., 2021; Mishra et al., 2024). Additionally, finger millet is also a good source of iron (4.6 mg/100 g), and comes second to barnyard millet. Moreover, it is another exceptional source of potassium, with its concentration being 137 mg/100 g (Rooney 1996). These concentrations of micronutrients are quite high when compared to other conventional sources of cereals, viz. rice and wheat. The majority of these micronutrients are reported to be found in the outer layers, i.e., the bran (Shobana et al. 2013). It has also been reported that various processing techniques can be useful in increasing the concentration of micronutrients in the finger millet grains. These include decortication, germination, soaking, fermentation, etc. and can enhance the concentrations significantly. A recent study was conducted by Arzoo et al. (2024) where they compared the impact of various processing techniques (soaking, parboiling, popping, germination) on the different properties of finger millet flour. The processing techniques had significantly reduced the concentration of tannins present in the sample. This meant that reduction in tannin may lead to an increase in the mineral bioavailability by reducing chelation. The reduction in tannin was in the following order: popping (0.59 mg/g) > soaking (0.46 mg/g) > germination (0.39 mg/g) > parboiling (0.22 mg/g). Another study focused on the influence of decortication, popping, malting on the bio-accessibility of Ca, Fe, and Zn in finger millet. It was found that decortication increased the bio-accessibility of Ca, Fe, and Zn; however, the total mineral content was decreased. Popping decreased Ca bio-accessibility but improved the Fe and Zn bio-accessibility. Malting showed the improvement of bio-accessibility of all the minerals (Krishnan, Dharmaraj, and Malleshi 2012). These studies show that presence of certain factors such as phytate concentration, presence of polyphenols, and interaction of dietary fiber can influence and significantly reduce the availability of minerals in finger millet, but the various conventional as well as non-conventional processing techniques can assist in the enhancement of mineral bioavailability (Sharma et al. 2021; Tripathi et al. 2014).

2.2.2 Antinutritional Factors

Millets are a diverse and nutritionally rich cereal grain which can provide various health benefits. However, as is the case with food products based from plant origin, millets are quite rich in antinutritional factors (ANFs) which can have an adverse effect if consumed in significant quantities. ANFs tend to form insoluble complexes, when ingested, by combining with the nutrients present in the host body and thereby reduce mineral bioavailability in the host. These can lead to mineral deficiencies of iron, zinc, etc. (Kumar, Tripathi et al. 2021). These ANFs include phytic acids, tannins, oxalic acids, and certain phenolics. Phytic acid/phytates are one of the major ANFs present in finger millet and other minor millets, which serve as the primary storage form of phosphorus. Minerals such as Ca^{2+}, Zn^{2+}, Mg^{2+}, etc. forms divalent cations with phytates which reduces mineral bioavailability in the body (Mishra, Kumar, and Srivastava 2024). Various cultivars of finger millet have also been reported to be rich in condensed tannins, which further interfere with millet digestibility. Various reports have shown that majority of the ANFs, along with phytochemicals, are distributed mostly in the seed coat of the grain (Anandharamakrishnan, Rawson, and Sunill 2022; Chethan and Malleshi 2007; Sharma et al. 2021). A study which investigated the biosynthesis pathways of oxalic acid in finger millet showed that these pathways modulated the calcium accumulation among the selected finger millet genotypes. They also conferred that formation of Ca-oxalate complex restricted calcium ion availability in the host, and thereby led to its accumulation (Akbar et al. 2018; Kumar, Tripathi et al. 2021).

2.2.3 Major Phytochemicals of Finger Millet

Millets are also rich in phytochemical content having potential bioactivities. These phytochemicals are the secondary metabolites found in millets, with flavonoids, phenolic acids, polyphenols being the major phytochemicals found in millets (Kumar, Rani et al. 2021). In finger millet, darker (brown) varieties have been reported to have higher phenolic activity when compared to the lighter varieties. In most cases, the seed coat has the highest concentration of phytochemicals which include various other forms that include carotenes, inhibitors, and other organic acids which may have numerous health benefits (Shobana et al. 2013; Kumar, Rani et al. 2021). In finger millet, the major phytochemicals with highest concentration are the phenolic acids and flavonoids. Additionally, concentrations of flavan-3-units (catechin, epigallocatechin) have also been reported in some finger millet varieties (Onipe and Ramashia 2022; Shobana et al. 2013). These phytochemicals may possess various antiaging, anti-diabetic, and antioxidative properties, as various research studies have reported, which play a crucial role in reduction of cancer and aging (Kumar, Rani et al. 2021). Since majority of these phytochemicals are found in the seed coat, Onipe and Ramashia (2022) investigated the nutritional properties as well as concentrations of phytochemicals, particularly arabinoxylan and polyphenols, in the finger millet seed coat (FMSC). The insoluble dietary fiber concentration was around 38.40–51% of total seed coat which included pectin, lignin, cellulose, and arabinoxylan. Arabinoxylan has various health benefits including diabetes treatment and can be used as prebiotics as well as a functional food ingredient. Polyphenols in the FMSC were found to be higher (2.3 g/100 g) than the whole meal of finger millet. However, studies are still required to explore the untapped potential of finger millet and the phytochemicals present, which can contribute toward providing health benefits as well as reduce food waste. Table 2.2 presents the comparative assessment of the phytochemical profile of finger millet with other millets.

TABLE 2.2
Phytochemical Profile of Finger Millet in Comparison with Other Millets

Millets	TPC	Caffeic acid (µg/g)	Ferulic Acid (µg/g)	Gallic Acid (µg/g)	Protocatechuic Acid (µg/g)
Finger	150–300 mg FAE/100 g	9.0–16.0	50–400	4.0–12.0	25.0–100.0
Foxtail	25–60 mg RE/100 g	12.0–40.0	225.0–850.0	4.0–7.0	10.0–15.0
Proso	2–30 mg GAE/100 g	15.0–340.0	150–400	4.0–7.0	60.0–75.0
Kodo	50–60 mg GAE/100 g	300.0–330.0	2210.0	2.0–4.0	60.0–70.0
Barnyard	40–50 mg GAE/100 g	8.10–12.0	170.0–215.0	2.0–4.5	-
Pearl	260–430 mg GAE/100 g	8.0–30.0	25.0–799.0	5.1–40.0	2.0–20.0

TPC—total phenolic content, FAE—ferulic acid equivalent, RE—rutin equivalent, GAE—gallic acid equivalent.

Source: Xiang et al. (2019), Pradeep and Sreerama (2017), Serna-Saldivar and Espinosa-Ramírez (2018), Dey et al. (2022), Kumar, Tripathi, et al. (2021), Anal et al. (2024), and Sharma et al. (2021).

2.3 Processing of Finger Millet for Nutritional Enhancement and Its Significance over Other Millets

Millets are still one of the most underutilized cereal grains despite its nutritional significance and being rich in micronutrients as well as phytochemicals. One of the major concerns is the presence of antinutritional compounds and its shelf-life significance. Finger millet is usually utilized after it has been decorticated for the preparation of various products such as *rotis* (Indian flat bread), *kanji* (porridge), composite flours, extruded products, noodles, gruels, etc. and various other traditional as well as commercial products (Jindal and Nikhanj 2023). Therefore, various conventional as well as novel processing approaches which have been applied to enhance the nutritional properties, improve the sensory properties and convenience, and improve the nutrient bioavailability in finger millet as well as among other millet grains. A major advantage in using millets for product development is that millets are gluten-free. Gluten-free products have seen massive rise in their market value in the recent years, and with finger millet being a major crop grown in most countries, proper processing techniques can introduce novel millet-based products and can aid in the upliftment of the general health and well-being of the population (Selladurai et al. 2023). Some conventional processing techniques include soaking and cooking, germination, malting, fermentation, decortication, popping, etc. In contrast, novel technologies involve thermal techniques such as microwave, infrared, radiofrequency, and ohmic heating and non-thermal techniques like cold plasma, pulse electric field, high-pressure processing, ultrasound treatment, etc. (Joshi, Singh, and Rao 2023; Saleh et al. 2013). Thus, processing is done primarily due to two reasons:

- Provide shelf-life stability to the millet products while preserving their nutritive qualities (Joshi, Singh, and Rao 2023; Yousaf et al. 2021b).
- Reduce the enzymatic activity of rancidity-causing enzymes, which include lipase, oxygenase, peroxidase, polyphenol oxidase (Joshi, Singh, and Rao 2023; Nanje Gowda et al. 2022).

Finger millet has exceptional malting characteristics, which can provide good quality malt for use in millet-based products. Moreover, due to its smaller grain size, even germination occurs which can help in favorable malt formation. This malt has the potential to be used in the preparation of weaning foods, and can be a suitable alternative to maltodextrin (Abioye, Ogunlakin, and Taiwo 2018; Shobana et al. 2013). Germination has been shown to be more effective in the reduction of antinutrients, and can further enhance the nutritional quality when combined with soaking, milling, and fermentation (Selladurai et al. 2023). A typical millet grain consists of the pericarp, seed coat, aleurone layer, endosperm, and germ layer (Anandharamakrishnan, Rawson, and Sunil 2022; Nanje Gowda et al. 2022). Milling is another conventional approach, where the finger millet kernel undergoes pulverization which separates the endosperm and seed coat of the kernel, and is further processed into millet flour. The whole finger millet flour is usually darker in color, which may affect its acceptability. Hence, improvements to lighten its color have been undergone to improve its acceptability. However, milling can lead to reduction in the total dietary fiber levels, with a considerable reduction in polished grain flours. On the contrary, milling may enhance the mineral absorption of finger millet flour, as levels of antinutrients are usually low in dehulled grains (Anandharamakrishnan, Rawson, and Sunil 2022). Various studies have revealed that milling has a significant impact on the physicochemical properties of finger millet. A study on hydrothermally treated finger millet grains showed that steeping (10 h, 30°C) followed by steaming at 98.0°C for 30 min reduced sugar concentration in the grain, with slight variations in amylose fraction (Selladurai et al. 2023). Another study by Dharshini and Meera (2023) investigated the influence of two processing techniques, popping and milling, on physicochemical properties of finger millet grown in Udhagamandalam. It was found that popping and milling, in combination, had led to a decrease in water activity (A_w), bulk density, and a reduction in the free-flowing capacity. However, there was an enhancement in the color of the grains. Textural differences due to structural changes were also observed due to milling in the form of agglomeration of damaged starch molecules on the surface of the grain. Fermentation also assists in enhancing the shelf-life of products, by inhibiting growth of spoilage molds and bacteria, as well as provide antifungal activity. Furthermore, lipolytic organisms present assist in glycerol and fatty acid breakdown, thereby improving the aroma, texture, odor, and taste in the fermented samples. Another study by Jan et al. (2022) subjected finger millet flour to various fermentation treatments using *Lactobacillus brevis, Lactobacillus plantarum, Saccharomyces cerevisiae, Saccharomyces cerevisiae* with $(NH_4)_2SO_4$, and with a combination of yeast and *L. brevis*. All the treatments significantly enhanced the phenolic, antioxidant, crude protein content, and mineral content, whereas antinutrients (phytates and tannins) were reduced. Overall, fermentation has the potential to enhance the protein and starch digestibility, along with decreasing the ANFs such as tannins, and significant favorable changes in the bioactive compounds of finger millet and its flour. Fermentation can also enhance the solubility, color, swelling power of finger millet grain and flour. These modifications can be suitable in processing value-added products from finger millet flour and can be used as functional food ingredients.

2.4 Future Perspectives

Millets are a nutritious alternative and widely consumed cereals but it need a greater promotion for the same level of popularity and acceptance as other major grains. Finger millet has high significant properties over other millet species due to its low glycemic index, high calcium content, rich nutritional profile, and longer shelf life. All these components make it ideal for future prospects of human health such as malnutrition, osteoporosis, and diabetes. As global health trends incline toward sustainable and nutrient-dense foods. Finger millet has great potential due to its high resilience to climate change and its ability to develop in marginal soil positions. Current trends indicate an upward interest in traditional and indigenous grains, driven by both consumer health consciousness and sustainability concerns. The rapid rise in demand for gluten-free food products provides a novel avenue for processing finger millet-based products and makes an initiation for consumer preference. To capitalize on this opportunity, where it has been predicted that the millet market will be a 12 million USD industry (Selladurai et al. 2023), producers need to rapidly ramp up technologies for improvement of finger millet processing techniques, which can reduce antinutrients drastically, enhance shelf-life, and provide nutritionally superior product to the consumers. However, finger millet faces stiff competition from other high-yielding crops and must find the way of market dynamics that often favor more widely known grains. Due to uncertainty in the global market, poor market assurance, and availability of cash crops, the cost-to-profit ratio is currently against the production of millet crops. Addressing these challenges while knowing its nutritional and environmental benefits will be key to unlocking finger millet as a potential crop in the future agricultural landscape. Moreover, creating awareness among consumers regarding the nutritional and health significance of millet will allow it to be eventually known and people will try to be more engaged with this nutritious grain.

Research in the field of biofortification and improved crop varieties and precision farming cultivation practices can enhance the yield of finger millet and its nutritional quality to make it a more accessible cereal. Processing techniques that can reduce the lipase activity and improve shelf-life require further research which can eventually provide products with superior nutritional and sensory qualities. Innovative techniques of food processing may expand the market for finger millet-based products and boost its gastronomic appeal. Another novel process that requires additional research engagement is millet by-product utilization, particularly the seed coat, or husk. The seed coat is a significant storehouse of various polyphenols, flavonoids, bioactive components, fiber, and has huge nutraceutical potential. Extraction of these compounds and their application for novel food product formulation will eventually be the key to more engagement and increased scientific curiosity on finger millet, as well as for other millet varieties.

2.5 Conclusion

Finger millet (*Eleusine coracana* L. Gaertn) is a highly nutritious and underutilized cereal crop with significant potential to address global food security challenges. It has a rich profile of proteins, carbohydrates, and dietary fiber, and presence of essential micronutrients makes it an excellent alternative to become staple grains like rice and wheat, particularly for underprivileged communities in Africa and Asia. The gluten-free nature of finger millet, combined with its bioactive compounds, positions it as a functional food that supports health outcomes such as diabetes management. Finger millet's rich content of bioactive compounds and phytochemicals positions it as a functional food that can contribute to better health outcomes, including the management of diabetes and improved gut health. Millets are a healthy choice that can help meet the growing food needs of people around the world in an uncertain environment. As the global population continues to rise and environmental challenges intensify, the cultivation and consumption of finger millet could play a crucial role in promoting sustainable agriculture and enhancing food security.

Processing techniques like germination, fermentation, and milling enhance nutrient bioavailability while reducing ANFs. Despite facing competition from more popular grains, the rising demand for gluten-free and nutrient-dense foods offers a unique market opportunity. Future research should focus on improving processing technologies, enhancing yields, and utilizing millet by-products. By increasing consumer awareness and integrating finger millet into mainstream diets, we can promote sustainable agriculture and contribute to a healthier food system. Future research and development efforts should focus on improving the accessibility and utilization of finger millet, supporting smallholder farmers, and raising awareness about its health benefits. By integrating finger millet into mainstream diets, we can take significant steps toward ensuring a more resilient and nutritious food system for all.

References

Abioye, V. F., G. O. Babarinde, G. O. Ogunlakin, J. A. Adejuyitan, S. J. Olatunde, and A. O. Abioye. 2022. "Varietal and Processing Influence on Nutritional and Phytochemical Properties of Finger Millet: A Review." *Heliyon* 8 (12): e12310. https://doi.org/10.1016/J.HELIYON.2022.E12310

Abioye, V. F., Ogunlakin, G. O., and Taiwo, G. 2018. "Effect of Germination on Anti-Oxidant Activity, Total Phenols, Flavonoids and Anti-Nutritional Content of Finger Millet Flour." *Journal of Food Processing & Technology* 9 (2). https://doi.org/10.4172/2157-7110.1000719

Akbar, Naved, Supriya Gupta, Apoorv Tiwari, K. P. Singh, and Anil Kumar. 2018. "Characterization of Metabolic Network of Oxalic Acid Biosynthesis through RNA Seq Data Analysis of Developing Spikes of Finger Millet (*Eleusine Coracana*): Deciphering the Role of Key Genes Involved in Oxalate Formation in Relation to Grain Calcium Accumulation." *Gene* 649 (April): 40–49. https://doi.org/10.1016/j.gene.2018.01.071

Anal, Anil Kumar, Ranjit Singh, Daniel Rice, Kingkan Pongtong, Udipta Hazarika, Dipika Trivedi, and Sanjaya Karki. 2024. "Millets as Supergrains: A Holistic Approach for Sustainable and Healthy Food Product Development." *Sustainable Food Technology*. Royal Society of Chemistry. https://doi.org/10.1039/d4fb00047a

Anandharamakrishnan, Chinnaswamy, Ashish Rawson, and C. K. Sunil. 2022. Handbook of millets-processing, quality, and nutrition status. Springer. https://doi.org/10.1007/978-981-16-7224-8

Arzoo, Neeru, Adarsh Kumar Shukla, Surbhi Panwar, and Ashwani Kumar. 2024. "Effect of Processing Techniques on Functional, Antioxidants, and Structural Properties of Finger Millet (*Eleusine Coracana*) Flour." *Food and Humanity* 2 (May): 100305. https://doi.org/10.1016/j.foohum.2024.100305

Brhane, Haftom, Teklehaimanot Haileselassie, Kassahun Tesfaye, Rodomiro Ortiz, Cecilia Hammenhag, Kibrom B. Abreha, and Mulatu Geleta. 2022. "Novel GBS-Based SNP Markers for Finger Millet and Their Use in Genetic Diversity Analyses." *Frontiers in Genetics* 13: 848627. https://doi.org/10.3389/fgene.2022.848627

Chamoli, Vinay, Anoop Badoni, Naveen Bahuguna, and Naman Joshi. 2018. "Millet (Eluesine Coracana):-Nutritional Status, Health Benefits and Processing Status-A Review." *Journal of Pharmacognosy and Phytochemistry* 5: 80–83.

Chethan, S, and G N Malleshi. 2007. "Finger Millet Polyphenols: Characterization and Their Nutraceutical Potential." *American Journal of Food Technology* 2 (7): 582–92.

Dey, Swarnima, Alok Saxena, Yogesh Kumar, Tanushree Maity, and Ayon Tarafdar. 2022. "Understanding the Antinutritional Factors and Bioactive Compounds of Kodo Millet (*Paspalum Scrobiculatum*) and Little Millet (*Panicum Sumatrense*)." *Journal of Food Quality*. https://doi.org/10.1155/2022/1578448

Dhanushkodi, V, A Thanga Hemavathy, S Shenbagavalli, S Sangeetha, R Anitha, and T UmaMaheshwari. 2023. "A Review on Nutritional Properties and Health Benefits of Finger Millet." *International Journal of Plant & Soil Science* 35 (18): 753–61. https://doi.org/10.9734/ijpss/2023/v35i183342

Dharshini, S. Sai, and M. Meera. 2023. "Effect of Popping and Milling on Physical, Chemical, Structural, Thermal Properties and Angle of Repose of Amaranth Seed (*Amaranthus Cruentus* L.) and Finger Millet (*Eleusine Coracana* L. Gaertn) from Udhagamandalam." *Applied Food Research* 3 (2). https://doi.org/10.1016/j.afres.2023.100306

Dida, Mathews M, and Katrien M Devos. 2006. "Finger Millet." In *Cereals and Millets*, 1st ed., 1: 333. Springer-Verlag Berlin. https://doi.org/10.1007/978-3-540-34389-9_10

Jan, Sumaira, Krishan Kumar, Ajar Nath Yadav, Naseer Ahmed, Priyanka Thakur, Divya Chauhan, Qurat Ul Eain Hyder Rizvi, and Harcharan Singh Dhaliwal. 2022. "Effect of Diverse Fermentation Treatments on Nutritional Composition, Bioactive Components, and Anti-Nutritional Factors of Finger Millet (*Eleusine Coracana* L.)." *Journal of Applied Biology and Biotechnology* 10: 46–52. https://doi.org/10.7324/JABB.2022.10s107

Jindal, Priyanka, and Pooja Nikhanj. 2023. "A Review on Processing Technologies for Value Added Millet Products." *Journal of Food Process Engineering*. https://doi.org/10.1111/jfpe.14419

Joshi, T. Jayasree, Shagolshem Mukta Singh, and Pavuluri Srinivasa Rao. 2023. "Novel Thermal and Non-Thermal Millet Processing Technologies: Advances and Research Trends." *European Food Research and Technology*. Springer Science and Business Media Deutschland GmbH. https://doi.org/10.1007/s00217-023-04227-8

K., Muhammad Yasir, Aparajita Bhasin, Baby Devi Maibam, and Maanas Sharma. 2024. "Review on Finger Millet (*Eleusine Coracana* L.): Nutritional Composition, Modification, Its Effect on Physicochemical, Structural, Functional Properties, and Its Applications." *Journal of Food Composition and Analysis* 135: 106623. https://doi.org/10.1016/j.jfca.2024.106623

Kaur, Simardeep, Arti Kumari, Karishma Seem, Gurkanwal Kaur, Deepesh Kumar, Surbhi Verma, Naseeb Singh, et al. 2024. "Finger Millet (*Eleusine Coracana* L.): From Staple to Superfood—A Comprehensive Review on Nutritional, Bioactive, Industrial, and Climate Resilience Potential." *Planta* 260 (3): 75. https://doi.org/10.1007/s00425-024-04502-2

Krishnan, Rateesh, Usha Dharmaraj, and Nagappa G. Malleshi. 2012. "Influence of Decortication, Popping and Malting on Bioaccessibility of Calcium, Iron and Zinc in Finger Millet." *LWT* 48 (2): 169–74. https://doi.org/10.1016/j.lwt.2012.03.003

Kumar, Anil, Manoj Kumar Tripathi, Dinesh Joshi, and Vishnu Kumar. 2021. *Millets and Millet Technology. Millets and Millet Technology*. Springer Singapore. https://doi.org/10.1007/978-981-16-0676-2

Kumar, Anil, Madhu Rani, Shalini Mani, Pallavi Shah, Dev Bukhsh Singh, Himabindu Kudapa, and Rajeev K. Varshney. 2021. "Nutritional Significance and Antioxidant-Mediated Antiaging Effects of Finger Millet: Molecular Insights and Prospects." *Frontiers in Sustainable Food Systems* 5 (October): 1–17. https://doi.org/10.3389/fsufs.2021.684318

Mishra, Sweta, Shailesh Kumar, and C. R. Srivastava. 2024. *Genetic Improvement of Small Millets*. Vol. 1. Springer Nature Singapore. https://doi.org/10.1007/978-981-99-7232-6

Nanje Gowda, N. A., P. V. Kaliramesh Siliveru, Yogita Vara Prasad, B. P. Bhatt, Chennappa Netravati, and Gurikar. 2022. "Modern Processing of Indian Millets: A Perspective on Changes in Nutritional Properties." *Foods*. MDPI. https://doi.org/10.3390/foods11040499

Onipe, Oluwatoyin Oladayo, and Shonisani Eugenia Ramashia. 2022. "Finger Millet Seed Coat—A Functional Nutrient-Rich Cereal By-Product." *Molecules*. MDPI. https://doi.org/10.3390/molecules27227837

Pradeep, P. M., and Yadahally N. Sreerama. 2017. "Soluble and Bound Phenolics of Two Different Millet Genera and Their Milled Fractions: Comparative Evaluation of Antioxidant Properties and Inhibitory Effects on Starch Hydrolysing Enzyme Activities." *Journal of Functional Foods* 35 (August): 682–93. https://doi.org/10.1016/j.jff.2017.06.033

Rakkammal, Kasinathan, Theivanayagam Maharajan, R. N. Shriram, P. S. Jeevan Ram, Stanislaus Antony Ceasar, and Manikandan Ramesh. 2023. "Physiological, Biochemical and Molecular Responses of Finger Millet (*Eleusine Coracana*) Genotypes Exposed to Short-Term Drought Stress Induced by PEG-6000." *South African Journal of Botany* 155 (April): 45–59. https://doi.org/10.1016/J.SAJB.2023.01.053

Rasri, W., Simmi Ranjan Kumar, and Pornya Khampratueng. 2024. "Valorization of Ragi Husk and Its Utilization in the Food Packaging Industry." In *Agro-Wastes for Packaging Applications*, 83–104. CRC Press

Rooney, L. W. 1996. "Sorghum and Millets". In Cereal Grain Quality, 153–177. Springer Netherlands. https://doi.org/10.1007/978-94-009-1513-8_5.

Saleh, Ahmed S. M., Qing Zhang, Jing Chen, and Qun Shen. 2013. "Millet Grains: Nutritional Quality, Processing, and Potential Health Benefits." *Comprehensive Reviews in Food Science and Food Safety* 12 (3): 281–95. https://doi.org/10.1111/1541-4337.12012

Selladurai, Manivannan, Manoj Kumar Pulivarthi, Anu Suprabha Raj, Mehreen Iftikhar, P. V. Vara Prasad, and Kaliramesh Siliveru. 2023. "Considerations for Gluten Free Foods – Pearl and Finger Millet Processing and Market Demand." *Grain and Oil Science and Technology*. KeAi Communications Co. https://doi.org/10.1016/j.gaost.2022.11.003

Serna-Saldivar, Sergio O., and Johanan Espinosa-Ramírez. 2018. "Grain Structure and Grain Chemical Composition." In *Sorghum and Millets: Chemistry, Technology, and Nutritional Attributes*, 85–129. Elsevier. https://doi.org/10.1016/B978-0-12-811527-5.00005-8

Sharma, Rajan, Savita Sharma, B. N. Dar, and Baljit Singh. 2021. "Millets as Potential Nutri-Cereals: A Review of Nutrient Composition, Phytochemical Profile and Techno-Functionality." *International Journal of Food Science and Technology*. https://doi.org/10.1111/ijfs.15044

Shobana, S., K. Krishnaswamy, V. Sudha, N. G. Malleshi, R. M. Anjana, L. Palaniappan, and V. Mohan. 2013. "Finger Millet (Ragi, *Eleusine Coracana* L.). A Review of Its Nutritional Properties, Processing, and Plausible Health Benefits." In *Advances in Food and Nutrition Research*, 69: 1–39. Academic Press Inc. https://doi.org/10.1016/B978-0-12-410540-9.00001-6

Singh, P., & Raghuvanshi, R. S. (2012). Finger millet for food and nutritional security. African Journal of Food Science, 6 (4): 77–84. 10.5897/AJFSX11.010

Sridhar, Rudravarapu, and Gollamudi Lakshminarayana. 1994. "Contents of Total Lipids and Lipid Classes and Composition of Fatty Acids in Small Millets: Foxtail (*Setaria Italica*), Proso (*Panicum Miliaceum*) and Finger (*Eleusine Coracana*)." *American Association of Cereal Chemists* 71 (4): 355–59.

Tripathi, Jaya, Alka Gupta, Ranu Prasad, and Vinita Puranik. 2014. "Enhancing micronutrient content of beverage powder by incorporating malted finger millet." *Indian Journal of Community Health* 26 (2): 339–342.

Xiang, Jinle, Franklin B. Apea-Bah, Victoria U. Ndolo, Mangani C. Katundu, and Trust Beta. 2019. "Profile of Phenolic Compounds and Antioxidant Activity of Finger Millet Varieties." *Food Chemistry* 275 (March): 361–68. https://doi.org/10.1016/j.foodchem.2018.09.120

Yousaf, Laraib, Dianzhi Hou, Humna Liaqat, and Qun Shen. 2021. "Millet: A review of its nutritional and functional changes during processing." *Food Research International* 142: 110197. https://doi.org/10.1016/j.foodres.2021.110197

3

Nutritional Inhibitors in Finger Millet

Ronak Tanwar, Anil Panghal, Shiksha Jangra, and Navnidhi Chhikara

3.1 Introduction

Finger millet (*Eleusine coracana* L.), nutritional and bioactive composition contributes to its nutraceutical properties and various health benefits including anti-diabetic, antimicrobial, and gluten-free characteristics (Gaikwad, 2024; Tanwar *et al.*, 2023). Components previously considered anti-nutrients, such as polyphenols, phytates, tannins, and dietary fiber, are now recognized for their antioxidant activity, combating aging, and metabolic diseases (Balyatanda Bheemaiah *et al.*, 2024) and accounts for diverse health advantages on finger millet consumption. Finger millet has three component namely bran, endosperm and germ. The red-brown seed coat contains different anti-nutritional compounds. Keeping in mind about health advantages of these bioactive compounds, adequate processing methods are selected. Various processing techniques, both traditional and modern (cooking, popping, steaming, parboiling, and fermentation) are employed to remove/reduce anti-nutritional factors (ANFs), enhance nutrient bioavailability and digestibility (Gowda *et al.*, 2022). These methods improve enzymatic activity, alter microstructure, increase water and oil holding capacity and enhance nutrient bio-accessibility (Tanwar *et al.*, 2025). Specifically, fermentation increases antioxidants and dietary fiber while reducing anti-nutrients (Gowda *et al.*, 2022) and germination protects against fungal attack, improves nutritive value, increases amylase levels, and reduces anti-nutrients like tannins and phytic acid, thereby enhancing mineral bio-accessibility (Singh *et al.*, 2023). Popping gelatinizes starch, sterilizes the grain, creates a pleasant aroma, and improves protein and starch digestibility (Kaur *et al.*, 2024), while parboiling alters starch, protein, and phenolic acid profiles, impacting grain composition and digestion (Shukla *et al.*, 2024). All processing treatments alone or in conjugation are applied on finger millet grain/flour to enhance palatability, product development and eradicating/minimizing the ANFs. The present chapter describes about various types of anti-nutrient, ill effects and processing applied for their reduction.

 DOI: 10.1201/9781003535119-3

3.2 Nutritional Profile of Finger Millet

Finger millet exceptional nutritional profile distinguishes it from other millets and cereals, making it a valuable dietary component. Its notable calcium and iron content contribute to its significance in balanced, nutrient-rich, and therapeutic diets. While pearl millet may exceed finger millet in protein, also boasts superior calcium levels. Similarly, though barnyard millet has higher fiber, while, finger millet fiber content surpasses that of proso and foxtail millets, which share comparable protein and carbohydrate levels. These distinctions highlight finger millet unique nutritional contribution (Kumar *et al.*, 2016a). Jayawardana *et al.* (2019) analyzed various finger millet varieties showing minimal compositional differences. Total dietary fiber ranged from 13.01% to 13.79% slowly digestible starch from 43.38% to 49.15% and resistant starch from 3.75% to 4.58%. The "Ravi" variety exhibited the highest ash content at 3.22%. Average mineral content across varieties (mg/100 g) included: sodium (12.04), magnesium (141.78), potassium (407.15), calcium (345.62), iron (3.49), zinc (1.89), and phosphorus (331.07). This consistent nutritional composition across varieties further emphasizes finger millet value as a reliable source of essential nutrients (Mishra and Singh, 2024).

3.3 Anti-Nutrients

Anti-nutrients are the bioactive compounds that interfere with the absorption/metabolism of nutrients and therefore adversely affect the nutritional quality of food. However, in some cases like phenolics, tannins, flavonoids, these might be providing some health benefits or biological significance on consumption in adequate amount. But, their higher level may showcase adverse impact on health on prolonged consumption. Millets produce ANFs as a defense mechanism against predators and harsh environments. These factors can limit nutrient availability and affect predators attack. Finger millet contains several such factors, including tannins, non-starch polysaccharides (NSPs), protease inhibitors, oxalates, and phytates, all of which can impact nutrient digestibility (Kumari *et al.*, 2021).

3.3.1 Polyphenols

Among cereals, finger millet, distinguished by its dark brown seed coat, stands out for its particularly rich phenol content (102 mg/100 g) surpassing that of barley (16.4 mg/100 g), rice (2.51 mg/100 g), maize (2.91 mg/100 g), and wheat (20.5 mg/100 g) (Devi *et al.*, 2014). Finger millet is a rich source of polyphenols, with over 50 identified compounds including phenolic acids, flavonols, flavones, and flavanonols, contributing to its high antioxidant capacity. These compounds exist in both soluble and bound forms, concentrated in the pericarp and cell walls of the kernel. Polyphenols are classified based on their carbon structure into groups like simple phenols, phenolic acids, tannins, flavonoids, lignans, lignins, curcuminoids, coumarins, and stilbenes. Studies have shown finger millet varieties to possess significant polyphenol content, impacting their antioxidant properties and potential health benefits (Hassan *et al.*, 2020). Studies have investigated distribution of polyphenols within millet seeds using histochemical and chemical analyses of different milling fractions. These analyses reveal that polyphenols are not uniformly distributed throughout grain but are concentrated in outer layers the aleurone layer, testa, and pericarp, which constitute the bran fraction. Histochemical examination suggests that approximately 60% of millet polyphenols are located in seed coat tissue, which comprises about 12% of seeds mass (Chethan & Malleshi, 2007). Phenolic compounds in grains exist in various forms: free, soluble conjugates, and insoluble bound forms. While Isaiah (2020) proposed that most phenolic compounds in millet are glycosides, Subba Rao and Muralikrishna (2002) identified ferulic acid (18.60 mg/100 g) as major bound phenolic acid and protocatechuic acid (45.0 mg/100 g) as major free phenolic acid. Ferulic acid and p-coumaric acid are primary bound phenolics in finger millet, representing 64–96% and 50–99% of total ferulic acid and p-coumaric acid content, respectively (Rajkumar *et al.*, 2024). Acidic methanol (1% HCl in methanol) has proven effective in extracting millet polyphenols. These polyphenols are heat-stable but sensitive to pH changes, particularly unstable under alkaline conditions (Chethan & Malleshi, 2007). High-performance liquid chromatography fractionation of extracted polyphenols has identified derivatives of benzoic acid (gallic acid, p-hydroxybenzoic acid and protocatechuic acid,) and cinnamic acid (syringic acid, p-coumaric acid, ferulic acid, and trans-cinnamic acid) along with flavonoid quercetin. Benzoic acid derivatives constitute about 85% of total phenolic compounds (Kuruburu *et al.*, 2022).Further analysis using ionization mass spectrometry of seed coat extract revealed the presence of additional compounds, including naringenin, kaempferol, luteolin glycoside, phloroglucinol, apigenin, (+)-catechin/(–)-epicatechin, trans-feruloyl-malic acid, a prodelphinidin dimer (epi/gallocatechins; 2GC), diadzein, catechin gallates, and trimers and tetramers of catechin (Onipe & Ramashia, 2022). This research highlights the complexity and diversity of phenolic compounds present in finger millet particularly within seed coat. The identification and quantification of these compounds contribute to a deeper understanding of the potential health benefits associated with finger millet consumption. Finger millet varieties contain proanthocyanidins, also called condensed tannins (Chandrasekara & Shahidi, 2011). These high-molecular-weight polyphenols, composed of polymerized flavan-3-ol and/or flavan-3,4-diol units possess biological activity. However, in high concentrations, they can reduce nutritional value and bioavailability of proteins and minerals. Condensed tannins possess beneficial properties like anti-inflammatory, antiviral, antibacterial, and antioxidant effects, being more potent antioxidants than their monomer counterparts. Among millets, finger millet exhibits the highest proanthocyanidin content 311.28±3.0 μmol of catechin equivalent/g of defatted meal exceeding levels found in other millets (finger millet, foxtail millet, little millet, pearl millet, and proso millet) and barley (Jayawardana *et al.*, 2021).

Polyphenols are increasingly recognized for their significant contributions to health and physiological function throughout the lifespan. Polyphenols, characterized by the presence of two or more phenolic groups, are also commonly marketed as dietary supplements. In millet predominant polyphenols are phenolic acids and tannins, with flavonoids present in lesser quantities. While not classified as essential nutrients, many polyphenols exhibit a wide range of beneficial properties, including antioxidant, anti-mutagenic, anti-estrogenic, anti-carcinogenic, anti-inflammatory, antiviral, and platelet aggregation inhibitory activities. These properties suggest a potential role for polyphenols in disease prevention and mitigation (Brglez Mojzer *et al.*, 2016).The dietary polyphenols can also interfere with thiamine and folic acid transport, alter drug activity, and inhibit iron absorption, potentially leading to iron deficiency. At high doses, some polyphenols may exhibit carcinogenic or genotoxic effects.

3.3.2 Tannins

Tannins, the fourth most abundant plant constituent after cellulose, hemicellulose, and lignin, are polyphenolic biomolecules recognized for their astringent taste and antioxidant properties. Classified into hydrolysable and condensed types, tannins are widely distributed across numerous plant families, varying according to origin. While tannins exhibit multifunctional properties in human body, their presence in food can negatively impact nutritional value. Tannins bind with proteins, reducing food efficiency, hindering growth, and impairing iron absorption. Excessive consumption can lead to gastrointestinal damage and depletion of essential tissues, proteins, and amino acids. The reducing tannin levels in food to safe levels for human consumption are crucial. In animals, tannins negatively affect feed intake, metabolizable energy, feed efficiency, growth rate, and protein digestibility across various species. Tannins are categorized as either hydrolysable, producing gallic acid upon hydrolysis, or condensed, yielding flavanols. Finger millet exhibits a wide range of tannin content (0.04–3.74% catechin equivalents), influenced by factors like grain color and variety. White grain varieties typically have lower tannin levels (0.05%) compared to brown and dark brown varieties (0.61%). Two African varieties IE927 and IE929 have been reported to contain the highest tannin amounts (3.42–3.47%). Dark-colored ragi varieties generally exhibit higher tannin levels, while brown varieties range from 0.35% to 2.40%, and white ragi contains negligible amounts (Wadikar *et al.*, 2007). Siwela *et al.* (2007) compared presence of tannins in 20-finger millets of white and brown types. It was observed that lighter coloured finger millet had quite lower tannin concentration compared to pigmented one and poor antioxidant activity therefore. This variation in tannin content across finger millet varieties has significant implications for nutritional value and potential applications.

3.3.3 Enzyme Inhibitors

Finger millet contains various enzyme inhibitors like protease, alpha-amylase, and trypsin inhibitors, which can affect digestion and other physiological processes. Protein inhibitors hinder protein digestion by blocking digestive enzymes trypsin and chymotrypsin, leading to pancreatic hypertrophy. Millets exhibit varietal differences in proteinase inhibitor activity, with finger millet showing higher trypsin inhibition activity than anti-chymotryptic activity (Veerabhadrappa *et al.*, 1978). The activity of protease inhibitors is substantially influenced by the pH of matrix. It was observed that under neutral conditions, the protease inhibitors exhibited thermal sensitivity and were inactivated, on the contrary, under acidic conditions exhibited thermal stability. The research also suggests comparable chymotrypsin inhibitory activity among different finger millets. Two distinct inhibitors, a chymotrypsin inhibitor and a trypsin alpha-amylase inhibitor, have been isolated from ragi using affinity techniques (Kumar *et al.*, 2016b).

Trypsin inhibitors, in particular, hinder protein digestion by binding to trypsin proteins. Protease inhibitors have diverse roles in plants, from storage protein functions to defense mechanisms against pests and pathogens. Their presence in finger millet can slow digestion and reduce the availability of certain amino acids. They can also influence biological processes in mammals. Reducing their content in food is therefore important. Amylase inhibitors, found in finger millet, target and inactivate alpha-amylases, further impacting carbohydrate digestion. Finger millet also contains other ANFs like tannins, NSPs, phytates, and oxalates, which can further affect nutrient digestibility. Studies have isolated and characterized trypsin inhibitors from finger millet, confirming their presence and potential impact on nutritional value (Bello *et al.*, 2022).

3.3.4 Non-Starch Polysaccharides

Besides tannins and protease inhibitors millets contain other ANFs like NSPs, phytates, and oxalates, which can impact nutrient digestibility. Finger millet also has slightly higher unavailable carbohydrates (18.6%) compared to wheat (17.3%) (Nirmala *et al.*, 2000). Finger millet pentosan content is reported as 6.2–7.2%, though native millets generally have more hexoses than pentosans. NSPs are polymeric carbohydrates distinct from starch in structure and composition, containing different glycosidic bonds that affect their breakdown by avian enzymes. These high-molecular-weight NSPs (8,000–9,000 million) can hinder broiler growth and feed efficiency (Senthilkumar *et al.*, 2025).

3.3.5 Phytate

Phytic acid (phytate) a phosphorus storage compound in plants is indigestible by humans and hinders mineral absorption. Comprising 1–5% of plant weight, it serves as a nutrient reserve in seeds. Its negative charge binds to positive mineral ions like calcium, magnesium, iron, and zinc, reducing their bioavailability. Finger millet found to contain 570 mg/100 g of phytic acid and is limiting factor for its staple usage (Shukla *et al.*, 2024). Phytate levels in millets fluctuate due to various factors, including growing conditions, processing, and testing methods. It's a major inhibitor of mineral absorption, prompting development of biofortified varieties to mitigate its effects (Singh *et al.*, 2017).

3.3.6 Oxalates

Finger millet has substantially higher total oxalates (11.3 mg/100 g) with lower soluble oxalates (45.9 mg/100 g) and 54.1% of insoluble oxalates (Amalraj & Pius, 2015). Soluble oxalate, which makes up the majority of the oxalate portion in the food matrix, can bind calcium and decrease its absorption. Therefore, finger millet has a higher percentage of bioavailable calcium (28%) than rice (24.7%), maize (25.4%), and sorghum (26%), while having higher levels of total oxalate than many other grains (Amalraj & Pius, 2015). The presence of oxalate in finger millet results in digestive issues, kidney stones, and hindrance in calcium absorption.

3.4 Strategies to Reduce Anti-Nutrients

Finger millet, or *Eleusine coracana*, is a staple grain in many parts of the world owing to its high nutritional content, especially the high levels of antioxidants, dietary fiber, and calcium. However, the presence of ANFs such oxalates, tannins, phytates, and enzyme inhibitors usually limits its nutritional value. These substances hinder digestion and lower the bioavailability of nutrients. Finger millet can be used more effectively as a dietary staple by reducing or eliminating these ANFs using a variety of processing techniques.

3.4.1 Decortication

Decortication, or dehulling, is a process that removes the outer layer (pericarp) of grains, which can contain varying husk percentages (1.5–29.3%) in different millets. Traditionally done manually, decortication now utilizes machines like rice hullers. Soaking time and the extent of decortication significantly impact tannin and phytate reduction. Studies show that 75% decortication with 90 minutes of soaking can drastically reduce tannins (from 337.09 to 11.9 mg/100 g) and phytates (from 442.12 to 113 mg/100 g). Increased soaking time and decortication generally lead to lower tannin content. Phytate reduction during soaking is attributed to leaching (Dharmaraj *et al.*, 2013). Research indicates that decortication can reduce phytic acid by up to 53% and total polyphenolic content by up to 9% in pearl millet. Dehulling has been shown to significantly reduce anti-nutrients in the bran of pearl millet cultivars, including condensed tannins (Samtiya *et al.*, 2021). Recent studies demonstrate varying phytate reductions through decortication across different millet types: 39% in little millet, 23% in barnyard millet, 25% in kodo millet, and 12% in common millet. Removing the pericarp and aleurone layer effectively lowers ANFs, making decortication a suitable method for eliminating phytates and tannins (Anitha *et al.*, 2024).

3.4.2 Heating

Heat treatments like roasting, boiling, cooking, and autoclaving effectively reduce anti-nutrients. Roasting significantly decreased tannins, phytates, trypsin inhibitors, and protease inhibitors by 74.6%, 28.4%, 98.3%, and 97.5%, respectively, while cooking reduced them by 42%, 75.8%, 95.8%, and 95.8%, respectively (Dhliwayo *et al.*, 2023). While some studies report increased total polyphenolic content after cooking, others show reductions in phytic acid and polyphenols. Popping process substantially reduced tannins (870.8–610.2 mg/100 g), phytic acid (851.4–333.1 mg/100 g), oxalic acid (45.8–32.2 mg/100 g) and trypsin inhibitors (4188–3090 U/g) in finger millet flour (Chauhan, 2018). Different heating methods, including microwave cooking, significantly decrease trypsin inhibitors, tannins, and phytic acid.

3.4.3 Soaking

Soaking is a commonly adopted method to reduce ANFs in various millets, legumes and pulses. Endogenous enzymes like phytase are activated when grains absorb water during soaking, breaking down phytates into phosphates and inositol. Soaking also releases oxalates and water-soluble tannins into the soaking media. According to research studies, soaking finger millet for 8–12 hours at room temperature can lower tannins by up to 15% and phytates by 20–30% (Rathore *et al.*, 2019; Samtiya *et al.*, 2020). However, further reduction can be enhanced by soaking in slightly acidic or alkaline solutions, including sodium bicarbonate or lemon juice (Chauhan, 2018). Soaking, often a precursor to germination, cooking, or fermentation, is a simple method for reducing anti-nutrients in food. Soaking for 12–18 hours can effectively lower levels of soluble phytic acid and proteolytic enzyme inhibitors. Discarding the soaking water removes leached polyphenols and other anti-nutrients, including phytates and phytic acid. Other research indicates a 36.8% reduction in phytic acid and a 39.57% reduction in polyphenols. Soaking for 16 hours has been shown to reduce tannins, phytic acid, and trypsin inhibitors. On soaking, decline in phytic acid and trypsin inhibitors by 13.22% and 13.51% in finger millet was observed (Patel & Dutta, 2018).

3.4.4 Germination

Germination is a biochemical process substantially improving nutritional and phytochemical content and further bioavailability along with minimizing ANFs. The grains are submerged in water for 6–8 hours for textural softening and endogenous enzymes like phytase, protease, tannase, and amylase are activated converting the complex nutrients into simpler ones. Germination for 48–72 hours has been shown to greatly improve the digestibility of proteins and starch while reducing the amount of tannins by 40% and the phytate content by up to 60% (Acharya *et al.*, 2024; Chauhan, 2018). The increased antioxidant activity of germinated finger millet contributes to the diet's functional health advantages (Samtiya *et al.*, 2021). Additionally, the sensory qualities like flavor and texture are enhanced, which makes germinated millet appropriate for use in fermented drinks, porridges, and weaning foods. Germination also reduces phytate content in millet and sorghum. Combining germination with fermentation yields positive results, decreasing anti-nutrients due to polyphenol loss during soaking (Mane *et al.*, 2023). In finger millet, maximum phytic acid (64.76%) and trypsin inhibitor (71.02%) reduction was observed after 36 hours of germination at 37°C while increasing protein digestibility and mineral bioavailability (Patel & Dutta, 2018). Germination also reduces the viscosity and bulk density of finger millet. Sprouting for 48 hours at 30°C is considered optimal, as longer periods result in further dry matter loss without significant nutritional improvement.

3.4.5 Fermentation

Fermentation is a traditional process used commonly for preparation of fermented drinks and porridges. ANFs like phytates and tannins are hydrolyzed by microbes during fermentation, especially lactic acid bacteria. It has been found that fermentation can drastically increase the bioavailability of minerals, particularly iron and zinc, by reducing phytates and tannins by 50–70% (Jeyakumar & Lawrence, 2022; Samtiya *et al.*, 2021). By enhancing flavor, texture, and shelf life, fermentation not only lowers ANFs but also improves the finger millet products' sensory and functional qualities. Fermentation enhances the nutritional value and antioxidant properties of food by reducing anti-nutrients like phytic acid, thereby improving protein digestion and bioavailability. Reduction in phenolic compounds, condensed tannins and phytic acid by 41%, 35%, and 22–54% and increase in mineral bioavailability was observed in finger millet porridge (Gabaza *et al.*, 2018). Adugna *et al.* (2024) evaluated impact of fermentation (24–48 hours) and malting (24–72 hours) on nutritional and anti-nutrient in finger millet biscuits. It was observed that substantial decline in tannin from 1.21 mg/100 g to 0.78 mg/100 g in biscuits prepared from fermented flour and 0.85 mg/100 g in biscuits from malted flour. Similar decline in phytic acid was observed from 435.50 mg/100 g to 430.78 mg/100 g and 431.60 mg/100 g, respectively.

TABLE 3.1

Impact of Processing Techniques on Finger Millet Properties

Processing	Effects	References
Decortication	Processing reduces compounds that interfere with mineral absorption (polyphenolic pigments and phytates), while simultaneously increasing levels of essential minerals like calcium, iron, and zinc.	Sunil *et al.* (2024)
Heating	Heat processing of finger millet, including roasting, boiling, and autoclaving, significantly reduces anti-nutrients like tannins, phytates, and enzyme inhibitors. The extent of reduction varies depending on specific heat treatment method and its parameters (time, temperature).	Abioye *et al.* (2022)
Soaking	Soaking finger millet involves steeping grains in water, changing water to prevent microbial growth, and then drying and milling. This process affects physical properties (size, density, color), as well as nutritional and functional characteristics (protein, fat, phytic acid, water absorption)	Sood *et al.* (2016)
Germination	Germination of finger millet reduces anti-nutrients like phytates and tannins, improving mineral bioavailability and protein digestibility. It also impacts physical properties like viscosity and bulk density, with optimal sprouting time around 48 hours at 30°C to balance nutrient enhancement and dry matter loss.	Chauhan (2018)
Fermentation	Fermentation of finger millet breaks down starch and long-chain fatty acids, reduces phytic acid and fat content, and increases bioavailability of minerals like iron, manganese, and calcium. It also enhances sensory characteristics (aroma, taste, flavor) and shelf life of final product.	Gaikwad (2024)
Popping or puffing	Popping finger millet involves heating grains to gelatinize starch and expand endosperm, which enhances digestibility, improves sensory qualities (color, taste, aroma), and reduces anti-nutrients. Different popping methods exist, including sand/salt treatment, hot air popping, and microwave heating, each impacting nutrient content differently.	Nanje *et al.* (2024)
Malting	Malting of finger millet reduces anti-nutritional factors (oxalates, phytates, tannins), increases nutrient content (vitamins, minerals, essential amino acids, carbohydrates), and improves bioavailability of iron and manganese. It also alters polyphenol content and enhances digestibility, with drying process being crucial for enzymatic activity, color, and flavor development.	Adugna *et al.* (2024)
Roasting	Roasting finger millet increases carbohydrate and ash content, enhances mineral bioavailability (calcium and iron), while reducing moisture, fat, protein, phenols, and antioxidant activity. It also increases resistant starch content, which is opposite to effect of other processing methods.	Singh *et al.* (2018)
Extrusion	Extrusion of finger millet uses high pressure and mechanical shear to create food products, reducing phytates and tannins while increasing mineral bioavailability. This process also affects viscosity of product (higher cold paste viscosity, lower cooked paste viscosity) and improves protein digestibility.	Devi *et al.* (2014)

3.4.6 Extrusion

Extrusion is a common high-temperature, quick processing method in industrial food production. It degrades ANFs by rupturing their structure with a combination of heat and mechanical shear. It has been demonstrated that extrusion cooking improves the texture and sensory appeal of finger millet products while reducing phytates that by 60–80% and tannins by 40% (Acharya *et al.*, 2024; Jeyakumar & Lawrence, 2022). Products made from extruded finger millet, like breakfast cereals and snacks, are becoming more and more well-liked because of their improved nutritional profile, extended shelf life, and consumers acceptability. Extrusion is the favored technique for large-scale manufacturing because of the quick processing time, which reduces nutritional losses.

3.4.7 Baking

Baking is dry heat cooking process converting dough into a baked product like bread, cookies, cakes etc. This procedure enhances the finished product's texture, flavor, and appearance while denaturing heat-sensitive ANFs like tannins and phytates. Baking works especially well for making bread, biscuits, and other baked products made with finger millet. Baking has been demonstrated to improve the digestibility of proteins and carbohydrates while reducing phytates by 20% and tannins by 35% (Chauhan, 2018; Rathore *et al.*, 2019). Baking is a great processing technique for contemporary food applications since it not only lowers ANFs but also maintains the nutritious content of finger millet, especially its dietary fiber and antioxidant qualities.

3.4.8 Boiling

One hydrothermal technique that successfully lowers water-soluble ANFs in finger millet is boiling. While heat denatures thermolabile substances like trypsin inhibitors, boiling causes tannins and oxalates to seep into the cooking water. It has been demonstrated that boiling for 15–30 minutes reduces phytates by 75%, oxalates by around 30%, and tannins by 42% (Rathore *et al.*, 2019). Boiling works well, but it can also lose some nutrients, especially water-soluble vitamins and minerals, thus cooking times and substrate water ratio must be carefully adjusted. However, boiled finger millet is frequently used in soups, porridges, and other traditional foods.

3.4.9 Roasting

Finger millet grains are roasted by applying dry heat, which lowers ANFs and improves the grain's flavor, aroma, and shelf life. High roasting temperatures denature heat-labile anti-nutritional substances such phytates, tannins, and enzyme inhibitors. Research indicates that roasting can lower phytates by 25% and tannins by up to 40% (Jeyakumar & Lawrence, 2022; Rathore *et al.*, 2019). The sensory appeal of finger millet can be enhanced with this process, which makes it acceptable for use in traditional recipes, flours, and snacks. Furthermore, it has been demonstrated that roasting increases the availability of specific antioxidants, which contributes to the functional health benefits of roasted millet products.

3.5 Conclusion

Finger millet stands out as a nutritional powerhouse, offering a wide range of health benefits. It is packed with essential nutrients like calcium, fiber, and bioactive compounds, including polyphenols and tannins, which contribute to its antioxidant, antimicrobial, and anti-diabetic properties. Its low glycemic index makes it an excellent choice for managing blood sugar levels and promoting overall health. While finger millet contains some compounds that can hinder nutrient absorption, these can be effectively reduced through traditional and modern processing methods. Techniques like decortication, germination, and fermentation not only minimize these anti-nutrients but also enhance the bioavailability of essential vitamins and minerals. Beyond its direct nutritional contributions, finger millet versatility extends to various applications. As a gluten-free grain, it serves as a valuable ingredient in a wide range of food products. Moreover, its byproducts find use in sustainable packaging and bioenergy production, contributing to a circular economy and reducing reliance on fossil fuels. Finger millet potential goes beyond its nutritional value, offering sustainable solutions for diverse industries. Its rich nutritional profile, combined with its versatility and sustainability, makes it a important grain for future. Further research and development are crucial to fully explore and utilize the multifaceted nature of finger millet, integrating it more effectively into modern diets and sustainable practices. This will not only enhance individual health and well-being but also contribute to a more sustainable and environmentally friendly future.

References

Abioye, V. F., G. O. Babarinde, G. O. Ogunlakin, J. A. Adejuyitan, S. J. Olatunde, and A. O. Abioye. "Varietal and processing influence on nutritional and phytochemical properties of finger millet: a review." *Heliyon* 8, no. 12 (2022): e12310.

Acharya, P., S. Karki, K. Rai, and A. Sangroula. "Effects of extrusion cooking over roasting on retention of nutritional quality of composite flour prepared from, millet, corn and rice blend." *Acta Universitatis Cibiniensis. Series E: Food Technology* 27, no. 2 (2024): 253–260.

Adugna, Ayele Assefa, Getachew Neme Tolesa, Solomon Abera, and Girma Daba Deme. "Effect of grain fermentation and malting time on nutrient and anti nutrient composition of biscuits from Aksum finger millet (*Eleusine coracana*)." *Cogent Food & Agriculture* 10, no. 1 (2024): 2336690.

Amalraj, Augustine, and Anitha Pius. "Influence of oxalate, phytate, tannin, dietary fiber, and cooking on calcium bioavailability of commonly consumed cereals and millets in India." *Cereal Chemistry* 92, no. 4 (2015): 389–394.

Anitha, Seetha, Priya Arjun, Nagarekha C. Palli, N. Sreekanth, S. A. Miruthika Devi, Sangeeta Pandey, Sridhar Krishnan, et al. "Sensory and nutritional evaluation of nine types of millet substituted for polished white rice in select Indian meal preparations." *Frontiers in Sustainable Food Systems* 8 (2024): 1331260.

Bello, Oluwasesan M., Abiodun B. Ogbesejana, Abdulrahman Balkisu, Martin Osibemhe, Bashir Musa, and Stephen O. Oguntoye. "Polyphenolic fractions from three millet types (Fonio, Finger millet, and Pearl millet): their characterization and biological importance." *Clinical Complementary Medicine and Pharmacology* 2, no. 1 (2022): 100020.

Bheemaiah Balyatanda, Suhan, N. A. Nanje Gowda, Jeyamkondan Subbiah, Snehasis Chakraborty, P. V. Vara Prasad, and Kaliramesh Siliveru. "Physiochemical, bio, thermal, and non-thermal processing of major and minor millets: a comprehensive review on antinutritional and antioxidant properties." *Foods* 13, no. 22 (2024): 3684.

Brglez Mojzer, Eva, Maša Knez Hrnčič, Mojca Škerget, Željko Knez, and Urban Bren. "Polyphenols: extraction methods, antioxidative action, bioavailability and anticarcinogenic effects." *Molecules* 21, no. 7 (2016): 901.

Chandrasekara, Anoma, and Fereidoon Shahidi. "Determination of antioxidant activity in free and hydrolyzed fractions of millet grains and characterization of their phenolic profiles by HPLC-DAD-ESI-MSn." *Journal of Functional Foods* 3, no. 3 (2011): 144–158.

Chauhan, E.S. "Effects of processing (germination and popping) on the nutritional and anti-nutritional properties of finger millet (*Eleusine coracana*)." *Current Research in Nutrition and Food Science Journal* 6, no. 2 (2018): 566–572.

Chethan, S., and N. G. Malleshi. "Finger millet polyphenols: optimization of extraction and the effect of pH on their stability." *Food Chemistry* 105, no. 2 (2007): 862–870.

Devi, Palanisamy Bruntha, Rajendran Vijayabharathi, Sathyaseelan Sathyabama, Nagappa Gurusiddappa Malleshi, and Venkatesan Brindha Priyadarisini. "Health benefits of finger millet (*Eleusine coracana* L.) polyphenols and dietary fiber: a review." *Journal of Food Science and Technology* 51 (2014): 1021–1040.

Dharmaraj, Usha, Ramasamy Ravi, and Nagappa G. Malleshi. "Optimization of process parameters for decortication of finger millet through response surface methodology." *Food and Bioprocess Technology* 6 (2013): 207–216.

Dhliwayo, T., P. Chopera, T. M. Matsungo, C. Chidewe, S. Mukanganyama, E. Nyakudya, F. Mtambanengwe, P. Mapfumo, and L. K. Nyanga. "Effect of germination and roasting on the proximate, mineral and anti-nutritional factors in finger millet (*Eleucine coracana*), cowpeas." *African Journal of Food, Agriculture, Nutrition and Development* 23, no. 8 (2023): 24346–24362.

Gabaza, Molly, Habtu Shumoy, Lindsey Louwagie, Maud Muchuweti, Peter Vandamme, Gijs Du Laing, and Katleen Raes. "Traditional fermentation and cooking of finger millet: implications on mineral binders and subsequent bioaccessibility." *Journal of Food Composition and Analysis* 68 (2018): 87–94.

Gaikwad, Vaibhav. "Nutritional significance of finger millet and its potential for using in functional products." *Foods and Raw Materials* 12, no. 1 (2024): 110–123.
Gowda, N. A. Nanje, Kaliramesh Siliveru, P. V. Vara Prasad, Yogita Bhatt, B. P. Netravati, and Chennappa Gurikar. "Modern processing of Indian millets: a perspective on changes in nutritional properties." *Foods* 11, no. 4 (2022): 499.
Hassan, Zahra Mohammed, Nthabiseng Amenda Sebola, and Monnye Mabelebele. "Assessment of the phenolic compounds of pearl and finger millets obtained from South Africa and Zimbabwe." *Food Science & Nutrition* 8, no. 9 (2020): 4888–4896.
Isaiah, O. O. (2020). "The synergistic interaction of phenolic compounds in pearl millets with respect to antioxidant and antimicrobial properties." *American Journal of Food Science and Health, 6,* 80–88.
Jayawardana, Sooriya Arachchige Sachini, Jayanetti Koralalage Ramani Radhika Samarasekera, Gardhi Hettiarachchige Chamari Madhu Hettiarachchi, Jaanaki Gooneratne, Saikat Datta Mazumdar, and Roopa Banerjee. "Dietary fibers, starch fractions and nutritional composition of finger millet varieties cultivated in Sri Lanka." *Journal of Food Composition and Analysis* 82 (2019): 103249.
Jayawardana, Sooriya Arachchige Sachini, Jayanetti Koralalage Ramani Radhika Samarasekera, Gardhi Hettiarachchige Chamari Madhu Hettiarachchi, Jaanaki Gooneratne, Muhammad Iqbal Choudhary, and Almas Jabeen. "Anti-inflammatory and antioxidant properties of finger millet (*Eleusine coracana* (L.) Gaertn.) varieties cultivated in Sri Lanka." *BioMed Research International* 2021, no. 1 (2021): 7744961.
Jeyakumar, Ebenezer, and Rubina Lawrence. "Microbial fermentation for reduction of antinutritional factors." In *Current developments in biotechnology and bioengineering*, pp. 239–260. Elsevier, 2022.
Kaur, Simardeep, Arti Kumari, Karishma Seem, Gurkanwal Kaur, Deepesh Kumar, Surbhi Verma, Naseeb Singh, et al. "Finger millet (*Eleusine coracana* L.): from staple to superfood—a comprehensive review on nutritional, bioactive, industrial, and climate resilience potential." *Planta* 260, no. 3 (2024): 75.
Kumar, Anil, Mamta Metwal, Sanveen Kaur, Atul K. Gupta, Swati Puranik, Sadhna Singh, Manoj Singh, et al. "Nutraceutical value of finger millet [*Eleusine coracana* (L.) Gaertn.], and their improvement using omics approaches." *Frontiers in Plant Science* 7 (2016a): 934.
Kumar, S. I., C. G. Babu, V. C. Reddy, and B. Swathi. "Anti-nutritional factors in finger millet." (2016b): 491.
Kumari, Parveen, Priyanka Kajla, and Shrestha Naudiyal. "Finger millet (*Eleusine coracana* L.)—properties and health benefits." *Handbook of Cereals, Pulses, Roots, and Tubers* (2021): 157–172.
Kuruburu, Mahadevaswamy G., Venugopal R. Bovilla, Rimshia Naaz, Zonunsiami Leihang, and SubbaRao V. Madhunapantula. "Variations in the anticancer activity of free and bound phenolics of finger millet (*Eleusine coracana* (L) Gaertn; variety KMR-301) seeds." *Phytomedicine Plus* 2, no. 2 (2022): 100276.
Mane, R. P., R. B. Kshirsagar, B. S. Agarkar, B. M. Patil, and S. D. Katke. "Effects of germination on antinutritional, colour, functional and physicochemical properties of millets." *Pharma Innov* 12, no. 1 (2023): 350–355.
Mishra, Sunita, and Rajani Singh. "A comprehensive study of nutritional profile, antioxidant activity and phytochemical properties of indigenous Indian millets." *African Journal of Food, Agriculture, Nutrition & Development* 24, no. 12 (2024).

Nanje Gowda, N. A., Vivek Kambhampati, Manoj K. Pulivarthi, Ruchi Chauhan, Ravi Pandiselvam, and Asgar Farahnaky. "Thermal and non-thermal bioprocessing: a comprehensive review on millet starch properties and digestibility." *Journal of Food Measurement and Characterization* (2024): 1–27.

Nirmala, M., M. V. S. S. T. Subba Rao, and G. Muralikrishna. "Carbohydrates and their degrading enzymes from native and malted finger millet (Ragi, *Eleusine coracana*, Indaf-15)." *Food Chemistry* 69, no. 2 (2000): 175–180.

Onipe, O. O., and Ramashia, S. E. (2022). Finger millet seed coat—A functional nutrient-rich cereal by-product. *Molecules, 27*(22), 7837.

Patel, Sumiran, and Samit Dutta. "Effect of soaking and germination on anti-nutritional factors of garden cress, wheat and finger millet." *International Journal of Pure and Applied Bioscience* 6, no. 5 (2018): 1076–1081.

Rajkumar, Haritha, Ilamaran Muthu, Prabhaharan James, Sivasankari Baskaran, Karpagapandi Lakshmanan, Meenakshi Vellaichamy, Selvi Jeyaraj, and Prema Pandi. "Enhancing nutritional profile, functional properties, therapeutic attributes of finger millet (*Eluesine coracana*) by germination: a comprehensive exploration." *Natural Product Communications* 19, no. 12 (2024): 1934578X241293027.

Rathore, Teena, Rakhi Singh, Dinkar B. Kamble, Ashutosh Upadhyay, and S. Thangalakshmi. "Review on finger millet: processing and value addition." *The Pharma Innovation Journal* 8, no. 4 (2019): 283–291.

Samtiya, Mrinal, Rotimi E. Aluko, and Tejpal Dhewa. "Plant food anti-nutritional factors and their reduction strategies: an overview." *Food Production, Processing and Nutrition* 2, no. 1 (2020): 14.

Samtiya, Mrinal, Komal Soni, Shashi Chawla, Amrita Poonia, Shalini Sehgal, and Tejpal Dhewa. "Key anti-nutrients of millet and their reduction strategies: an overview." *Acta Scientific Nutritional Health* 5, no. 12 (2021): 2582–1423.

Senthilkumar, Shankar, Khumbaron Kiranbala Kabui, and K. A. Athmaselvi. "Anti-nutritional factors in millets and methods to mitigate them." *Sign* (2025). https://foodnbeveragesprocessing.com/anti-nutritional-factors-in-millets-and-methods-to-mitigate-them/ (assessed on August, 2025)

Shukla, Adarsh Kumar, Surbhi Panwar, and Ashwani Kumar. "Effect of processing techniques on functional, antioxidants, and structural properties of finger millet (*Eleusine coracana*) flour." *Food and Humanity* 2 (2024): 100305.

Singh, Ajay, Saurabh Gupta, Ramandeep Kaur, and H. R. Gupta. "Process optimization for anti-nutrient minimization of millets." *Asian Journal of Dairy and Food Research* 36, no. 4 (2017): 322–326.

Singh, Neha, John David, D. K. Thompkinson, Blessy Sagar Seelam, Hridesh Rajput, and Sonia Morya. Effect of roasting on functional and phytochemical constituents of finger millet (*Eleusine coracana* L.). *The Pharma Innovation Journal, 7*(4) (2018): 414–418.

Singh, Poornima, Vinay Kumar Pandey, Zainab Sultan, Rahul Singh, and Aamir Hussain Dar. "Classification, benefits, and applications of various anti-nutritional factors present in edible crops." *Journal of Agriculture and Food Research* 14 (2023): 100902.

Siwela, Muthulisi, John R. N. Taylor, Walter A. J. De Milliano, and Kwaku G. Duodu. "Occurrence and location of tannins in finger millet grain and antioxidant activity of different grain types." *Cereal chemistry* 84, no. 2 (2007): 169–174.

Sood, Salej, Anil Kumar, B. Kalyana Babu, Vikram S. Gaur, Dinesh Pandey, Lakshmi Kant, and Arunava Pattnayak. "Gene discovery and advances in finger millet [*Eleusine coracana* (L.) Gaertn.] genomics—an important nutri-cereal of future." *Frontiers in Plant Science* 7 (2016): 1634.

Subba Rao, M. V. S. S. T., and G. Muralikrishna. "Evaluation of the antioxidant properties of free and bound phenolic acids from native and malted finger millet (Ragi, *Eleusine coracana* Indaf-15)." *Journal of Agricultural and Food Chemistry* 50, no. 4 (2002): 889–892.

Sunil, C. K., Gowda, N. N., Nayak, N., and Rawson, A. (2024). Unveiling the effect of processing on bioactive compounds in millets: Implications for health benefits and risks. *Process Biochemistry, 138*, 79–96.

Tanwar, R., A. Panghal, A. Kumari, and N. Chhikara. "Effect of finger millet on nutritional, rheological, cooking and pasting profile of Asian noodles." *Journal of Food Chemistry & Nanotechnology* 9 (2023): S647–S652.

Tanwar, Ronak, Anil Panghal, Anju Kumari, and Navnidhi Chhikara. "Effect of hydrothermal treatments on physicochemical, rheological and nutritional properties of millet: an analytical review." *Grain & Oil Science and Technology* (2025).

Veerabhadrappa, Patnagere S., Nagenahally H. Manjunath, and Tumkur K. Virupaksha. "Proteinase inhibitors of finger millet (*Eleusine coracana Gaertn*)." *Journal of the Science of Food and Agriculture* 29, no. 4 (1978): 353–358.

Wadikar, D. D., Premavalli, K. S., Satyanarayanaswamy, Y. S., and Bawa, A. S. (2007).

"Lipid profile of finger millet (Eleusine coracana) varieties." *Journal of Food Science and Technology, 44*(1), 79–81

4

Bioactive Compounds: Structure, Properties, and Applications of Finger Millet

Amrita Poonia, Priya Gupta, and Pawan Kumar Singh

4.1 Introduction

Eleusine coracana (L.), often known as finger millet, is an allotetraploid (2n = 4X = 36, AABB) belonging to the Poaceae family and genus Eleusine. Finger millet is a self-pollinating crop with a 1,593 Mb genome. This annual herbaceous cereal crop is grown and consumed by impoverished people in Asia and Africa. It has more protein and nutrients than other essential grains, including wheat, rice, and sorghum (Antony Ceasar et al., 2018). Finger millet is known for having an exceptionally high calcium (Ca) content. The seeds are a good source of dietary fiber, iron, phytates, trypsin inhibitory factors, and essential amino acids (leucine, isoleucine, methionine, and phenylalanine). They don't contain gluten either. Furthermore, finger millet provides many health benefits, such as hypoglycemic, hypocholesterolemic, and anti-ulcerative qualities (Gebreyohannes et al., 2024). Finger millet grains are malted and used as a healthy baby food. They are also used as flour to make cakes, bread, and other pastries. Additionally, the sprouted seeds are nutrient-dense and easily digested. By 2050, it is anticipated that there will be approximately 9.8 billion people on the planet. By 2050, the world's food consumption is expected to reach 14,886 million tons. To address this, there will be a need to expand the output of finger millet from 2.5 to 4.5 t ha^{-1} to meet the demand for grains (Maharajan et al., 2022).

In Asia and Africa, where famine and malnutrition are pervasive, an increase in the production of finger millet with high nutritional content will be advantageous to the agricultural systems. Finger millet has a great nutritional profile, but it suffers from illnesses including blast, lodging, and moisture stress, a lack of better varieties, a lack of research attention, and difficulties in threshing and milling (Wambi et al., 2021).

The nutritional content, variety of health benefits, and use of finger millet in food products are highlighted in this chapter. Future studies in finger

DOI: 10.1201/9781003535119-4

millet and other millets would build on this chapter to better understand and improve nutrient transport.

4.2 Nutritional Profile of Finger Millet

4.2.1 Macronutrients

The high carbohydrate content of finger millet is mainly made up of dietary fiber and non-starchy polysaccharides. It has 72 g of carbohydrates per 100 g, of which 1.04% are free sugars and 11.5% are non-starchy polysaccharides. In finger millet, 80–85% amylopectin and 15–20% amylose make up the starch granules (Shingote et al., 2021; Rishitha & Soni, 2024). A balanced amino acid composition and 7.16 g/100 g of protein are found in finger millet. Methionine, tryptophan, and lysine are among the essential amino acids that are notably abundant in finger millet, even though cereals and millets typically contain less lysine than legumes and animal proteins. Other important amino acids, like as valine, threonine, isoleucine (4.3 g), leucine (10.8 g), methionine (3.1 g), and phenylalanine (6.0 g) are also present, making it one of the most nutritious millets. The low fat content of finger millet is one of its qualities, which lessens any potential health risks. Grain-based essential fatty acids, including linolenic acid (1.3–4.4 mg/100 g) and palmitic acid (21.1–24.7 mg/100 g), support the development of the brain and neural tissue. It also has other good fatty acids, such as oleic acid (49.8 mg/100 g) and linoleic acid (24.2 mg/100 g) (Ramashia et al., 2019; Thagunna et al., 2022).

4.2.2 Micronutrients

Finger millet has a 2.7% total mineral content, with calcium and other vital micronutrients, including iron, magnesium, sodium, phosphorus, copper, and potassium being particularly abundant. The mineral-rich areas are the pericarp, aleurone layer, and germ. Vitamin B complexes, such as thiamine, riboflavin, folic acid, and niacin, are also abundant in finger millet. Other vitamins found in finger millet include thiamine, niacin, riboflavin, and tocopherol. Fresh grains also contain vitamin C, though dried grains lack this vitamin (Jagati et al., 2021). Finger millet has a rich mineral profile, with significant quantities of calcium (2400.00–3400.00 mg kg^{-1}), iron (40.28–47.60 mg kg^{-1}), zinc (12.40–17.45 mg kg^{-1}), phosphorus (1600.00–2900.00 mg kg^{-1}), potassium (3800.00–5200.00 mg kg^{-1}), and manganese (51.33–61.28 mg kg^{-1}). The high calcium and iron content of this mineral-dense grain may help with several health problems, such as anemia and bone and dental abnormalities. Finger millet, a great source of vital nutrients, can help reduce bone-related health issues. Regular body processes and general health depend on these nutrients (Anitha et al., 2021; Nakarani et al., 2021).

4.2.3 Dietary Fiber

Finger millet is a rich source of soluble and insoluble dietary fibers. However, compared to other millets, finger millet has about 22% more dietary fiber (3.6 g per 100 g). The five-layered testa of the grain contributes to its high dietary fiber content. One of the factors contributing to the grain's high dietary fiber content is its five-layered testa. This fiber provides satiety while consuming fewer calories because it is bulky and digests slowly (Rishitha & Soni, 2024). Dietary fiber concentration varied among the finger millet cultivars Ravi, Rawana, and Oshadha; on a dry weight basis, soluble fiber ranged from 13.01% to 13.79%, insoluble fiber from 0.52% to 0.59%, and total dietary fiber from 12.49% to 13.20% (Jayawardana et al., 2019). These dietary fibers serve as essential elements for improving digestive health, regulating blood sugar, and maintaining healthy gut flora. Its high dietary fiber content makes finger millet a nutritious option to combat chronic illnesses like diabetes, heart disease, and some types of cancer (Singh et al., 2022).

4.3 Classification of Bioactive Compounds

Finger millet is rich in phytochemicals, also called bioactive compounds, which have several health benefits. The bioactive substances found in finger millet are classified as phenolic acids, flavonoids, phytosterols, and other phytochemicals (Mutshinyani et al., 2020). It has been confirmed that the phenolic acids included in finger millet, such as coumaric acid, caffeic acid, and ferulic acid, have anti-inflammatory and antioxidant properties. Additionally, finger millet includes flavonoids like quercetin and kaempferol, which have been linked to several health benefits, like cardiovascular health and anti-cancer properties. One of the phytosterols in finger millet that has been shown to have cholesterol-lowering qualities is beta-sitosterol. Due to its many different bioactive components, finger millet is a beneficial grain for improving health and well-being (Thagunna et al., 2022; Shrestha et al., 2024; Maja et al., 2025).

4.3.1 Polyphenols

One of the most explored groups of naturally occurring antioxidants is polyphenols. In terms of structure, they can be divided into several subclasses, such as lignans, stilbenes, flavonoids, phenolic acids, and other phenolic compounds, which are distinguished by the presence of an aromatic ring with at least one hydroxyl group (de Araújo et al., 2021). The wide range of health advantages of dietary plant polyphenols has attracted the attention of consumers, nutritionists, and medical researchers. In addition to possibly lessening the signs of aging, these advantages include lowering the risk of several chronic diseases, including diabetes, cancer, cardiovascular disease, neurological diseases, and infections (Kumar et al., 2016). About 60% of the polyphenol content in finger millet comes from its bran, making it a rich source of polyphenols. Phenolic acids and tannins constitute the majority of the polyphenol composition, whereas flavonoids comprise a smaller percentage. Phenolic substances found in finger millet encompass tannins, bound phenolics such ferulic acid and p-coumaric acid, and free phenolics like protocatechuic acid. The polyphenol concentration of finger millet has been found to vary by variety, with brown variants having a greater polyphenol content (1.2–2.3%) than white types (0.3–0.55%). Geographical differences in polyphenol concentration have also been observed; African types range from 0.54% to 3.4%, whereas Indian variants range from 0.08% to 0.96% (Kaur et al., 2024).

Scavenging reactive oxygen species, modifying signaling pathways, and blocking pro-inflammatory enzymes and cytokines comprise every part of mode of action of finger millet polyphenols. More specifically, these polyphenols have anti-inflammatory actions by inhibiting the synthesis of pro-inflammatory enzymes and cytokines. This is accomplished by altering important cellular pathways, such as the Nrf2 pathway, which controls antioxidant responses, and the NF-κB pathway, which controls inflammation. Additionally, the polyphenols in finger millet have been demonstrated to work in concert with other nutrients and bioactive substances to boost their antioxidant activity (Balkrishna et al., 2024).

Polyphenols are essential for preserving the natural balance between antioxidants and oxidants, enabling them to avoid oxidative stress. Polyphenols are strong antioxidants that have many health advantages, especially their anti-aging effects. Because of their importance in preserving health, polyphenols are increasingly considered to be crucial for extending life (Chandrasekara & Shahidi, 2010). Polyphenol content of some compounds has been associated to other beneficial effects, such as hypoglycemic, hypocholesterolemic, and anti-ulcerative properties, emphasizing their significance for overall well-being and health (Kumar et al., 2021).

4.3.2 Flavonoids

The variety of flavonoids found in finger millet includes chalcones, anthocyanins, flavonols, flavones, and aminophenol compounds. Approximately 2 mg/g of flavonoids are present in defatted finger millet food (Abioye et al., 2022). Four Malawian finger millet varieties (white, brown, reddish, and red) have total flavonoid values ranging from 90.24 to 202.94 mg catechin acid equivalent/100 g. The total flavonoid content of the brown variants was higher than that of the other categories. The white variant exhibited the lowest scores across all assessed antioxidant characteristics (Xiang et al., 2019). Proanthocyanidins are naturally occurring compounds found in millet grains that are oligomeric or polymeric flavonoids made up of flavan-3-ol molecules. It was discovered that proanthocyanidins were more prevalent in brown finger millets (0.12–3.47% catechin equivalents) than in white finger millets (0.04–0.06%) (Shahidi & Chandrasekara, 2013). It has been found that flavonols have antioxidant and anti-inflammatory effects, such as kaempferol and quercetin, may assist in preventing off chronic illnesses, including cancer and heart disease. Other flavanols that have been connected to better cardiovascular health and a lower risk of cancer are epicatechins and catechins. Because of their anti-inflammatory and antioxidant properties, flavanones like naringenin and hesperidin may help lower the incidence of neurodegenerative disorders (Kumar et al., 2022). Fruits and vegetables obtain their red, purple, and blue hues from anthocyanins, which have been demonstrated to have strong antioxidant and anti-inflammatory effects. They are attributed to improved cardiovascular health, a lower risk of several diseases, and improved cognitive function (Olawoye et al., 2024).

4.3.3 Phenolic Acids

Hydroxycinnamic and hydroxybenzoic acids are the two predominant forms of phenolic acids. Gallic acid, protocatechuic acid, and vanillic acid are examples of hydroxybenzoic acids that can be found in a wide range of fruits, vegetables, and cereals. However, hydroxycinnamic acids like ferulic acid, caffeic acid, and syringic acid are frequently found in whole grains, fruits, and vegetables (Gull et al., 2014). Finger millet extracts primarily contained ferulic acid, gallic acid, caffeic acid, syringic acid, and chlorogenic acid at 57.04±0.28, 11.93±0.11, 38.91±0.15, 30.05±0.52, and 2.29±0.07 for the free phenolic fraction (FPF), while 56.12±0.31, 762.12±1.26, 31.86±0.20, 0 and 5.41±0.31 for the bound phenolic fraction (BPF) (Goudar et al., 2023). Phenol safeguards against age-related illnesses like type II diabetes, cardiovascular disease, obesity, and cancers, as well as diseases brought on by elevated oxidative stress, such as those caused by certain diets. Numerous metabolic diseases are linked to lifestyle choices, primarily a diet deficient in fiber and bioactive substances (Ashik Somarajan & Morya, 2022).

Gallic acid constitutes a trihydroxybenzoic acid which assists human keratinocytes in healing wounds (Yang et al., 2016). Syringic acid is a naturally produced phenolic compound that is considered to be a metabolite of plants. It is a 3,5-dimethyl ether derivative of gallic acid. Syringic acid is a phenolic chemical that has anti-inflammatory, anti-bacterial, anti-cancer, anti-diabetic, and antioxidant properties (Vo et al., 2020). The conjugated C3-side chain and phenolic nucleus of ferulic acid give it antioxidant properties that protect against oxidative damage and chronic illnesses (de Oliveira Silva & Batista, 2017). A flavoring compound made from dihydroxybenzoic acid, vanillic acid has nephroprotective actions against diabetes and antioxidant qualities. Anti-inflammatory, anti-cancer, and antioxidant properties are among the possible health advantages of these phenolic acids. They are crucial for fostering general health and well-being since they may help prevent and manage several illnesses (Morya, 2021).

4.3.4 Tannins

Finger millet has a greater tannin concentration than other millets, with a wide range of 0.04–3.47%. Finger millet cultivars vary in their tannin content; white versions have lower tannin levels (0.05%) than dark brown ones (0.61%). By lowering iron intake, a high tannin concentration might have a detrimental effect on the grain's nutritional composition. Furthermore, it has been demonstrated that in experimental animals, elevated tannin levels reduce growth rate, feed intake, protein digestibility, and net metabolizable energy (Kumar et al., 2016). The nutritional advantages of finger millet can be increased, and the detrimental effects of tannins can be lessened with the use of processing techniques and cultivar selection. A diet high in tannins harms intestinal digestion, cellulose levels, and microbial enzyme activity. Tannins use chemical neutralization to shield our bodies against cell harm. Tannins have beneficial effects on the cardiovascular system, antibacterial, and anti-cancer properties (Rathore et al., 2019; Jan et al., 2022).

4.3.5 Phytates and Antinutritional Factors

Finger millets contain phytic acid, a potent antinutritional component. The presence of additional antinutritional components in finger millets, such as oxalic acid, trypsin, and chymotrypsin, can also impact the absorption and utilization of nutrients. Phytic acid, a phosphorus-storing substance found in cereal grains, may have health effects because it can chelate vital minerals including magnesium, zinc, and iron. The phytic acid level of finger millet was 468±5.03 mg/100 g. Calcium and other minerals may be bound by oxalic acid, which will decrease the rate of absorption. The oxalic acid content of finger millet was 39.38 mg/100 g (Goudar et al., 2023). For those who depend on finger millet as a main diet, these antinutritional components can have serious consequences, especially in areas where mineral shortages are common. However, a variety of processing techniques can aid in lowering these antinutritional factors. The enzyme phytase, which degrades phytic acid, can be activated by fermentation, soaking, and germination, boosting the bioavailability of minerals. Tannins and oxalic acid levels can also be decreased by cooking and processing (Sheethal et al., 2022; Dhliwayo et al., 2023).

4.3.6 Bioactive Peptides

Bioactive peptides are byproducts of various biological sources that improve physiological conditions or functions, which in turn impact an individual's health. Therefore, bioactive peptides of millets have been found to have distinct biochemical effects on human health, such as antioxidant,

ACE-inhibitory, antidiabetic, antibacterial, anticancer, and antihypertensive characteristics (Ganguly et al., 2019; Majid & Priyadarshini, 2020). Finger millet proteins and peptides exhibited a wide spectrum of antibacterial activity against *Salmonella entrica* (MTCC 739) and *Pseudomonas aeruginosa* (MTCC 424) (Bisht et al., 2016).

4.3.7 Saponins and Other Secondary Metabolites

A secondary metabolite analysis was performed on three finger millet varieties: Khulule, Okhale-1, and U-15. Extracts from leaves and roots included tannins, glycosides, and phenols. While root extracts exhibited lower quantities of flavonoids, leaf extracts had higher levels of alkaloids and flavonoids. Both the leaf and the root extracts had different levels of terpenoids. Okhale-1 and U-15 root extracts had intermediate levels of saponins, while Ikhulule root extracts had significant levels. Secondary metabolite levels were generally moderate to high in leaf extracts and lower in root extracts (Chrisantus et al., 2024). Finger millet contains saponins (0.36%) with hemolytic action. Saponins are steroid glycosides or triterpenoid compounds with antibacterial, anti-inflammatory, and antioxidant properties. It has an impact on both the central nervous system and the gastrointestinal tract. Additionally, it results in cardiovascular disease. Protein digestion and absorption are reduced by saponin. Saponins may assist with reducing cholesterol and post-meal blood sugar levels as well as potentially prevent cancer (Kumar et al., 2016; Rathore et al., 2019).

4.3.8 Impact of Processing on Bioactive Compounds

Finger millet grains can be processed using both conventional and innovative techniques to help change them into edible foods. Soaking, germination, malting, fermentation, milling or grinding, heating, roasting, popping or puffing, and other traditional grain processing techniques are generally employed in rural areas. The traditional processing of finger millet food products has been poorly applied scientifically, especially in developing countries (Saleh et al., 2013; Subastri et al., 2015). Micronutrient bioavailability is improved by soaking finger millet grains in distilled water for 30°C to 60°C for the whole night. Grain is cleaned and dried at 60°C for 90 minutes after discarding the soaking water. This procedure raises the availability of zinc while decreasing antinutritional elements like phytic acid. The nutritional value of finger millet grains is increased by soaking, which increases their suitability for consumption. This is a quick and easy solution to improve finger millet flour's nutritional profile (Banusha & Vasantharuba, 2013).

Finger millet grains are soaked and then incubated for 24–48 hours at 30°C in order to achieve conventional germination. This inexpensive method improves amino acids, minerals, and carbs while softening grain structure and improving nutritional value. Grain functionality has been enhanced by germination for generations without the need for costly machinery. It is a quick and easy way to improve the nutritional value of finger millet (Shimray et al., 2012). Malting is a process that reduces antinutrients and improves nutritional content by steeping, germination, drying, toasting, grinding, and sieving. While phytic acid and tannins are reduced, fiber, vitamins, and minerals are improved. Malting makes a product more nutrient-dense and easily digestible by improving the sensory qualities and digestibility of starch (Thapliyal & Singh, 2015; Ramashia et al., 2019).

A popular method of processing finger millet grains is fermentation, which acts as a growing medium for microorganisms. Typically, it takes 24–72 hours for preparing traditional foods and beverages. In Africa, fermentation is commonly employed to produce well-liked native fermented goods and improves nutritional content (Rathore et al., 2019; Mutshinyani et al., 2020). The microbial burden is decreased and the sensory quality is improved by boiling, mashing, and combining finger millet grains with water to produce a soup. An alternative is to prepare a porridge by combining flour with hot water. These cooking techniques produce traditional and wholesome meals by enhancing the grain's texture and safety for ingestion (Ramashia et al., 2019).

4.4 Health Benefits of Finger Millet Bioactives

An absence of nutritional supplements and foods high in protein leads to protein-energy malnutrition, particularly for children below five and mothers who breastfeeding in developing countries like India. According to the World Health Organization, four primary forms of malnutrition affect people globally. This includes iron, iodine, vitamin A, and protein energy deficiency. Because millets are a storehouse of nutrients, they can aid with malnutrition, which affects the vast majority of our people (Jagati et al., 2021; Dhanushkodi et al., 2023). Finger millet has numerous nutritional advantages, such as controlling blood sugar and turning it into insulin, preventing diabetes with high dietary fiber and polyphenols, strengthening bones with high dietary fiber and polyphenols, acting as an anti-aging agent with antioxidants and essential amino acids, maintaining blood pressure levels with high fiber and low glycemic index, preventing diseases with anti-inflammatory and anticancer properties, improving hemoglobin status with iron and calcium, acting as a natural probiotic for diarrhea, improving lactation with amino acids, calcium, and iron, and offering unique therapeutic effects with antioxidative, anti-inflammatory, and antimicrobial properties and antioxidative properties (Srivastava & Sharma, 2012; Devi et al., 2014; Abah et al., 2020; Dhanushkodi et al., 2023; Vagdevi et al., 2023).

4.5 Applications in Functional Foods and Nutraceuticals

Finger millets have drawn interest because of their high nutritional profile and bioactive compounds, which could potentially use in functional meals and dietary supplements. Functional meals made from finger millet might include snack bars, baked items, and drinks that are high in fiber, vitamins, and minerals (Banerjee et al., 2012). The use of gluten-free cereals has significantly increased as a result of consumers refusing gluten-containing goods. The gluten-sensitive society has switched to diets that include gluten-free or low-gluten (20 parts per million) grains. Worldwide utilization of gluten-free products is projected to continue to rise at a 7.6% annual rate during 2020 and 2027 (Selladurai et al., 2023). Finger millet is a multipurpose cereal grain used in both conventional and novel ways to produce a wide range of culinary products. These products fall into two main categories: unconventional food items, which include baked goods, extruded products, popped snacks, and specialized beverages, and conventional food items, which include pancakes, flatbreads, porridges, and fermented products. The variety of products made from finger millet demonstrates the potential as a useful ingredient in several kinds of industrial and dietary applications (Kubo, 2016; Kumar et al., 2020; Kaur et al., 2024).

Furthermore, the lignocellulosic substances present in Finger millet straw and husk show great potential for the production of synthetic polymer composites for a variety of applications, such as food packaging. These agricultural wastes are used as reinforcement to make low-cost thermoplastic polymers with better elastic modulus, together with polypropylene, polyethylene, polylactic acid, polyester acetate, polyvinyl acetate, and other materials (Haque et al., 2022; Kaur et al., 2024).

The starch from native finger millet, an underutilized resource in India, was utilized for manufacturing of biodegradable films with enhanced mechanical and thermal properties. By adding sorbitol as a plasticizer, the characteristics of film were much improved, indicating possible uses in the manufacturing of environmentally acceptable and sustainable edible coatings and packaging alternatives (Gautam et al., 2022).

4.6 Conclusion

With its high dietary fiber, antioxidant, and anti-inflammatory qualities, finger millet has a strong nutritional profile and a wide range of uses. These include preventing chronic diseases and supporting digestive, weight, and immune health. Its adaptability is further evidenced by its use in both conventional food products and cutting-edge biodegradable packaging materials, which demonstrates its potential to improve human health and environmental sustainability. Finger millet is an important crop that simultaneously supports agricultural progress, human well-being, and ecological sustainability due to its numerous benefits. Through the promotion of sustainable farming methods, its production and use can support local food systems and communities.

In conclusion, finger millet proves to be a useful crop with broad applications in sustainable agriculture, conservation of the environment, and human health. Research and development on the crop is appealing due to its potential uses in nutraceuticals, functional meals, and biodegradable packaging. The potential of finger millet to improve health outcomes, support sustainable agriculture, and contribute to a more ecologically conscious food system can be fully explored by encouraging its cultivation, applications, and improvement.

References

Abah, C. R., Ishiwu, C. N., Obiegbuna, J. E., & Oladejo, A. A. (2020). Nutritional composition, functional properties and food applications of millet grains. *Asian Food Science Journal, 14*(2), 9–19.

Abioye, V. F., Babarinde, G. O., Ogunlakin, G. O., Adejuyitan, J. A., Olatunde, S. J., & Abioye, A. O. (2022). Varietal and processing influence on nutritional and phytochemical properties of finger millet: A review. *Heliyon, 8*(12), 1–15.

Anitha, S., Givens, D. I., Botha, R., Kane-Potaka, J., Sulaiman, N. L. B., Tsusaka, T. W., & Bhandari, R. K. (2021). Calcium from finger millet—A systematic review and meta-analysis on calcium retention, bone resorption, and in vitro bioavailability. *Sustainability, 13*(16), 8677.

Antony Ceasar, S., Maharajan, T., Ajeesh Krishna, T. P., Ramakrishnan, M., Victor Roch, G., Satish, L., & Ignacimuthu, S. (2018). Finger millet [*Eleusine coracana* (L.) Gaertn.] Improvement: Current status and future interventions of whole genome sequence. *Frontiers in Plant Science, 9*, 1054.

Ashik Somarajan, K., & Morya, S. (2022). A review on finger millet properties, processing, health benefits, and applications. *The Pharma Innovation Journal, 11*(7), 3388–3395.

Balkrishna, A., Verma, S., Singh, S. K., Dobhal, V., & Arya, V. (2024). Exploring the antioxidant mechanisms of millet polyphenols: Regulation of Nrf2 and NF-κB and their impact on gut microbiota. *Discover Food, 4*(1), 123.

Banerjee, S., Sanjay, K. R., Chethan, S., & Malleshi, N. G. (2012). Finger millet (*Eleusine coracana*) polyphenols: Investigation of their antioxidant capacity and antimicrobial activity. *African Journal of Food Science, 6*(13), 362–374.

Banusha, S., & Vasantharuba, S. (2013). Effect of malting on nutritional contents of finger millet and mung bean. *American-Eurasian Journal of Agriculture and Environmental Science, 13*(12), 1642–1646.

Bisht, A., Thapliyal, M., & Singh, A. (2016). Screening and isolation of antibacterial proteins/peptides from seeds of millets. *International Journal of Current Pharmaceutical Research, 8*(3), 96–99.

Chandrasekara, A., & Shahidi, F. (2010). Content of insoluble bound phenolics in millets and their contribution to antioxidant capacity. *Journal of Agricultural and Food Chemistry, 58*(11), 6706–6714.

Chrisantus, K. M., Sarah, C., Dorcas, L., Ramkat, R. C., Oduori, C. O., & Pili, N. N. (2024). Characterization of finger millet extracts and evaluation of their nematicidal efficacy and plant growth promotion potential. *Plant-Environment Interactions, 5*(4), e70006.

de Araújo, F. F., de Paulo Farias, D., Neri-Numa, I. A., & Pastore, G. M. (2021). Polyphenols and their applications: An approach in food chemistry and innovation potential. *Food Chemistry, 338*, 127535.

de Oliveira Silva, E., & Batista, R. (2017). Ferulic acid and naturally occurring compounds bearing a feruloyl moiety: A review on their structures, occurrence, and potential health benefits. *Comprehensive Reviews in Food Science and Food Safety, 16*(4), 580–616.

Devi, P. B., Vijayabharathi, R., Sathyabama, S., Malleshi, N. G., & Priyadarisini, V. B. (2014). Health benefits of finger millet (*Eleusine coracana* L.) Polyphenols and dietary fiber: A review. *Journal of Food Science and Technology, 51*, 1021–1040.

Dhanushkodi, V., Hemavathy, A. T., Shenbagavalli, S., Sangeetha, S., Anitha, R., & UmaMaheshwari, T. (2023). A review on nutritional properties and health benefits of finger millet. *International Journal of Plant & Soil Science, 35*(18), 753–761.

Dhliwayo, T., Chopera, P., Matsungo, T. M., Chidewe, C., Mukanganyama, S., Nyakudya, E., ... & Nyanga, L. K. (2023). Effect of germination and roasting on the proximate, mineral and anti-nutritional factors in finger millet (*Eleucine coracana*), cowpeas. *African Journal of Food, Agriculture, Nutrition and Development, 23*(8), 24346–24362.

Ganguly, A., Sharma, K., & Majumder, K. (2019). Food-derived bioactive peptides and their role in ameliorating hypertension and associated cardiovascular diseases. *Advances in Food and Nutrition Research, 89*, 165–207.

Gautam, N., Garg, S., & Yadav, S. (2022). Development of flexible and thin films from underutilized Indian finger millet (*Eleusine coracana*) starch. *Starch-Stärke, 74*(3–4), 2100156.

Gebreyohannes, A., Shimelis, H., Mashilo, J., Odeny, D. A., Tadesse, T., & Ojiewo, C. O. (2024). Finger millet (*Eleusine coracana*) improvement: Challenges and prospects—A review. *Plant Breeding, 143*(3), 350–374.

Goudar, G., Manne, M., Sathisha, G. J., Sharma, P., Mokalla, T. R., Kumar, S. B., & Ziouzenkova, O. (2023). Phenolic, nutritional and molecular interaction study among different millet varieties. *Food Chemistry Advances, 2*, 100150.

Gull, A., Jan, R., Nayik, G. A., Prasad, K., & Kumar, P. (2014). Significance of finger millet in nutrition, health and value added products: A review. *Magnesium (mg), 130*(32), 120.

Jagati, P., Mahapatra, I., & Dash, D. (2021). Finger millet (Ragi) as an essential dietary supplement with key health benefits: A review. *International Journal of Home Science, 7*(2), 94–100.

Jan, S., Kumar, K., Yadav, A. N., Ahmed, N., Thakur, P., Chauhan, D., & Dhaliwal, H. S. (2022). Effect of diverse fermentation treatments on nutritional composition, bioactive components, and anti-nutritional factors of finger millet (*Eleusine coracana* L.). *Journal of Applied Biology and Biotechnology, 10*(1), 46–52.

Jayawardana, S. A. S., Samarasekera, J. K. R. R., Hettiarachchi, G. H. C. M., Gooneratne, J., Mazumdar, S. D., & Banerjee, R. (2019). Dietary fibers, starch fractions and nutritional composition of finger millet varieties cultivated in Sri Lanka. *Journal of Food Composition and Analysis, 82*, 103249.

Kaur, S., Kumari, A., Seem, K., Kaur, G., Kumar, D., Verma, S., … & Riar, A. (2024). Finger millet (*Eleusine coracana* L.): From staple to superfood—A comprehensive review on nutritional, bioactive, industrial, and climate resilience potential. *Planta, 260*(3), 75.

Kubo, R. (2016). The reason for the preferential use of finger millet (*Eleusine coracana*) in eastern African brewing. *Journal of the Institute of Brewing, 122*(1), 175–180.

Kumar, A., Kumari, P., & Kumar, M. (2022). Role of millets in disease prevention and health promotion. In *Functional foods and nutraceuticals in metabolic and non-communicable diseases* (pp. 341–357). Academic Press.

Kumar, A., Metwal, M., Kaur, S., Gupta, A. K., Puranik, S., Singh, S., … & Yadav, R. (2016). Nutraceutical value of finger millet [*Eleusine coracana* (L.) Gaertn.], and their improvement using omics approaches. *Frontiers in Plant Science, 7*, 934.

Kumar, A., Rani, M., Mani, S., Shah, P., Singh, D. B., Kudapa, H., & Varshney, R. K. (2021). Nutritional significance and antioxidant-mediated antiaging effects of finger millet: Molecular insights and prospects. *Frontiers in Sustainable Food Systems, 5*, 684318.

Kumar, P., Kaur, C., Sethi, S., & Kaur Jambh, H. (2020). Effect of extruded finger millet on dough rheology and functional quality of pearl millet-based unleavened flatbread. *Cereal Chemistry, 97*(5), 991–998.

Kumar, S. I., Babu, C. G., Reddy, V. C., & Swathi, B. (2016). Anti-nutritional factors in finger millet.

Maharajan, T., Ceasar, S. A., & Ajeesh Krishna, T. P. (2022). Finger millet (*Eleusine coracana* (L.) Gaertn): Nutritional importance and nutrient transporters. *Critical Reviews in Plant Sciences, 41*(1), 1–31.

Maja, D., Udeh, H. O., Maphosa, T. M., Bhebhe, E., & Gwata, E. T. (2025). Mineral constituents and bioactive compounds of finger millet grain. In *Food security and nutrition* (pp. 151–168). CRC Press.

Majid, A., & Priyadarshini, C. G., P. (2020). Millet derived bioactive peptides: A review on their functional properties and health benefits. *Critical Reviews in Food Science and Nutrition, 60*(19), 3342–3351.

Morya, J. K. S. (2021). Celiac disease: An epidemiological condition: Insight on gluten free diet, significance and regulatory recommendations. *The Pharma Innovation Journal, 10*(5), 641–654.

Mutshinyani, M., Mashau, M. E., & Jideani, A. I. O. (2020). Bioactive compounds, antioxidant activity and consumer acceptability of porridges of finger millet (*Eleusine coracana*) flours: Effects of spontaneous fermentation. *International Journal of Food Properties, 23*(1), 1692–1710.

Nakarani, U. M., Singh, D., Suthar, K. P., Karmakar, N., Faldu, P., & Patil, H. E. (2021). Nutritional and phytochemical profiling of nutracereal finger millet (*Eleusine coracana* L.) genotypes. *Food Chemistry, 341*, 128271.

Olawoye, B., Popoola-Akinola, O. O., Popoola, O. O., & Akanbi, C. T. (2024). Nutraceutical compounds and potential human health benefits of pigmented millets. In *Pigmented grains* (pp. 199–216). Academic Press.

Ramashia, S. E., Anyasi, T. A., Gwata, E. T., Meddows-Taylor, S., & Jideani, A. I. O. (2019). Processing, nutritional composition and health benefits of finger millet in Sub-Saharan Africa. *Food Science and Technology, 39*, 253–266.

Rathore, T., Singh, R., Kamble, D. B., Upadhyay, A., & Thangalakshmi, S. (2019). Review on finger millet: Processing and value addition. *The Pharma Innovation Journal, 8*(4), 283–291.

Rishitha, P., & Soni, D. (2024). Finger millet: Nutritional profile and potential health benefits. *Research Junior Agricultural Science, 15*(1), 153–158.

Saleh, A. S., Zhang, Q., Chen, J., & Shen, Q. (2013). Millet grains: Nutritional quality, processing, and potential health benefits. *Comprehensive Reviews in Food Science and Food Safety, 12*(3), 281–295.

Selladurai, M., Pulivarthi, M. K., Raj, A. S., Iftikhar, M., Prasad, P. V., & Siliveru, K. (2023). Considerations for gluten free foods-pearl and finger millet processing and market demand. *Grain & Oil Science and Technology, 6*(2), 59–70.

Shahidi, F., & Chandrasekara, A. (2013). Millet grain phenolics and their role in disease risk reduction and health promotion: A review. *Journal of Functional Foods, 5*(2), 570–581.

Sheethal, H. V., Baruah, C., Subhash, K., Ananthan, R., & Longvah, T. (2022). Insights of nutritional and anti-nutritional retention in traditionally processed millets. *Frontiers in Sustainable Food Systems, 5*, 735356.

Shimray, C. A., Gupta, S., & Venkateswara Rao, G. (2012). Effect of native and germinated finger millet flour on rheological and sensory characteristics of biscuits. *International Journal of Food Science and Technology, 47*(11), 2413–2420.

Shingote, A. B., Sadawarte, S. K., Pawar, V. S., & Gaikwad, K. K. (2021). Studies on chemical and mineral evaluation of raw rice, sorghum, ragi and green gram. *The Pharma International Journal, 10*, 337–340.

Shrestha, J., Subedi, S., Shrestha, S., Rai, S., Poudel, A., Adhikari, S., … & Karanjit, R. (2024). Crop biofortification of finger millet through agronomic approaches to enhance bioactive compounds for human health and nutrition. *Agronomy Journal of Nepal, 8*, 237–249.

Singh, V., Lee, G., Son, H., Amani, S., Baunthiyal, M., & Shin, J. H. (2022). Anti-diabetic prospects of dietary bio-actives of millets and the significance of the gut microbiota: A case of finger millet. *Frontiers in Nutrition, 9*, 1056445.

Srivastava, K., & Sharma, A. K. (2012). Nutraceutical importance of finger millet (*Eleusine coracana*) for improved human health. *European Journal of Plant Science Biotechnology, 6*, 91–5.

Subastri, A., Ramamurthy, C., Suyavaran, A., Mareeswaran, R., Mandal, P., Rellegadla, S., & Thirunavukkarasu, C. (2015). Nutrient profile of porridge made from *Eleusine coracana* (L.) grains: Effect of germination and fermentation. *Journal of Food Science and Technology, 52*, 6024–6030.

Thagunna, B., Rimal, A., Kaur, J., Dhakal, Y., & Paudel, B. (2022). Finger millet: A powerhouse of nutrients its amino acid, micronutrient profile, bioactive compounds, health benefits, and value-added products. *Journal of Research in Agriculture and Animal Science, 9*, 36–44.

Thapliyal, V., & Singh, K. (2015). Finger millet: Potential millet for food security and power house of nutrients. *International or Research in Agriculture and Forestry, 2*(2), 22–33.

Vagdevi, A., Trupthi, S., Pandey, M., Reddy, N. B., & Singh, M. (2023). Nutritional qualities, processing and health benefits of finger millet (*Eleusine coracana* L.). *The Pharma Innovation Journal, 12*(1), 1291–1297.

Vo, Q. V., Bay, M. V., Nam, P. C., Quang, D. T., Flavel, M., Hoa, N. T., & Mechler, A. (2020). Theoretical and experimental studies of the antioxidant and antinitrosant activity of syringic acid. *The Journal of organic chemistry, 85*(23), 15514–15520.

Wambi, W., Otienno, G., Tumwesigye, W., & Mulumba, J. (2021). Genetic and genomic resources for finger millet improvement: Opportunities for advancing climate-smart agriculture. *Journal of Crop Improvement, 35*(2), 204–233.

Xiang, J., Apea-Bah, F. B., Ndolo, V. U., Katundu, M. C., & Beta, T. (2019). Profile of phenolic compounds and antioxidant activity of finger millet varieties. *Food Chemistry, 275*, 361–368.

Yang, D. J., Moh, S. H., Son, D. H., You, S., Kinyua, A. W., Ko, C. M., … & Kim, K. W. (2016). Gallic acid promotes wound healing in normal and hyperglucidic conditions. *Molecules, 21*(7), 899.

5

Processing of Finger Millet

Angelika Nelson

5.1 Introduction

Finger millet (*Eleusine coracana*) is a highly nutritious and resilient crop, well-suited to diverse agro-climatic conditions. Its processing is crucial in enhancing its nutritional value, digestibility, and adaptability in various food applications. Traditional processing techniques, such as milling and malting, are still widely used, but modern methods like extrusion cooking and baking have expanded the range of products derived from finger millet. Milling involves grinding the grains to produce flour, which is commonly used for porridge or bread. Malting, on the other hand, involves soaking, germinating, and drying the grains, a process that boosts the bioavailability of essential nutrients like proteins and vitamins (Abulude et al., 2005). Extrusion cooking applies heat and pressure to create puffed or expanded products, which are convenient and nutrient-rich. Baking, another key process, transforms finger millet flour into ready-to-eat products like biscuits and cakes. Each of these methods not only improves the taste, texture, and shelf life of finger millet-based products but also enhances their health benefits, making them more appealing to consumers. This chapter will delve into the scientific principles behind these processing techniques, illustrating how they contribute to the creation of value-added products that meet the demands of a modern, health-conscious market. Finger millet (*Eleusine coracana*), is a vital crop known for its nutritional benefits and resilience in various agro-climatic conditions. Its processing involves multiple steps, each aimed at enhancing its edibility, digestibility, and usability in various food products. The processing techniques for finger millet range from traditional milling and malting to more advanced methods like extrusion cooking and baking ragi (Antony et al., 1996; Banusha & Vasantharuba, 2013). This chapter provides an in-depth review of these processes, highlighting their scientific underpinnings and practical applications. We will explore how these

DOI: 10.1201/9781003535119-5

methods transform finger millet from raw grains into versatile, value-added products.

5.2 Milling

Milling serves as the foundational processing step for finger millet, transforming harvested grains into flour. This process is of paramount importance as it directly influences the quality, texture, and overall characteristics of the flour, which subsequently dictates the properties of the final food products. Unlike some other cereals, finger millet presents unique challenges due to its kernel structure. The kernel comprises a fragile endosperm encased within an intact seed coat. This combination of a delicate interior and a robust outer layer prevents the grain from being polished and cooked in its whole form or as grits, similar to rice or other common cereals (Katti et al., 2008). Consequently, finger millet must invariably undergo pulverization or milling to produce flour suitable for consumption. The primary purpose of milling finger millet is to create flour, which is the base ingredient for most finger millet-based dishes. The milling process typically begins with a thorough cleaning stage. This crucial step removes any extraneous materials that may have been harvested along with the grains, such as stones, weed seeds, stalks, and other debris. This cleaning ensures the purity and safety of the final flour product. Following cleaning, the finger millet grains are passed through specialized milling equipment, commonly grating or grinding plants. These machines work to separate the glumes, which are the inedible cellulosic outer layers of the grain. Removing the glumes is essential as they contribute undesirable texture and do not offer nutritional value. After the removal of the glumes, the remaining portion of the grain is then pulverized or milled into fine flour. This pulverization is the defining characteristic of finger millet processing, as it is the only way to effectively utilize the grain for food preparation. The resulting flour is then ready for use in a wide variety of culinary applications, from traditional porridges and breads to more modern baked goods and snacks. Therefore, the conversion of finger millet grain into flour through pulverization or milling represents the most common and essential primary processing method for this versatile grain (Kurien & Desikachar, 1966).

Different types of conventional and modern equipment are used for this purpose: conventional stone mills; burr mills (steel or emery type); hammer mills; ball mills; and small pulverisers. The fineness of the flour depends on the type of mill used.

5.2.1 Pre-Processing Steps

Before milling, finger millet grains undergo the following pre-processing steps:

a. Cleaning and aspiration to remove impurities
b. Destoning to remove stones
c. Grading to sort grains by size
d. Degumming or surface scoring to remove the seed coat
e. Fumigation for disinfection
f. Bulk storage

5.2.2 Milling Process

There are three main types of milling processes:

5.2.2.1 Direct Milling

It involves roller milling, where the grains are passed through rollers to separate the endosperm and bran. This method produces flour with a high dietary fiber content and retains the natural reddish-brown to dark brown color of finger millet (Malleshi & Desikachar, 1981a). Direct milling, however, results in coarse flour with a significant amount of bran and germ, which can affect the flour's texture and color. A machine for milling has now been developed in India. This so-called "mini millet mill" consists of a water mixer, a plate grinder, and various sifter attachments. It is a versatile device in which debranning and sizing the endosperm (into either flour or semolina) take place in a single operation. It yields fairly white products. It can also be used to process wheat, maize, sorghum, and pearl millet and will even remove the outer husk from finger millet seeds if the clearance between the grinder plates is reduced (Kurien et al., 1959).

5.2.2.2 Hydrothermal Processing

This involves moistening, conditioning, drying, grinding, and sieving of the grains. Typically, finger millet is steeped in water for around 24 hours for 30°C and 70°C for 2 hours, steamed, dried to about 12–13% moisture, and then ground into flour. This method improves milling efficiency and hardness, producing finer flour suitable for various applications. Hydrothermal treatment also enhances the preparation of grits but impacts product quality due to increased hardness. The sequence of hydrothermal processing steps is illustrated in Figure 5.1. Hydrothermal processing involves the application of heat and moisture to the millet grains. This process typically includes steps such as soaking, steaming, or boiling, which serve to:

- *Gelatinize Starch*: Hydrothermal treatment modifies the starch structure, making it more digestible and improving its functional properties, such as solubility and swelling power.
- *Alter Grain Morphology*: The structure of the endosperm changes from a soft and fragile texture to a more coherent mass, which increases hardness and alters the grain's physical properties, such as density and sphericity.
- *Improve Cooking Qualities*: Hydrothermal processing enhances the cooking quality of finger millet, resulting in reduced peak viscosity and zero breakdown viscosity during cooking, which is beneficial for various culinary applications (Rajasekhar et al., 2018).

5.2.2.3 Decortication

Decortication in the milling of finger millet is a crucial process that involves the removal of the tough seed coat from the edible endosperm. This step is essential for improving the cooking quality and palatability of the millet, as the seed coat can negatively affect both. The debranning or decortication methods followed for most of the cereals were not effective in the case of finger millet owing to the intactness of the seed coat with highly fragile endosperm. Hence, to decorticate, finger millet is hydrothermally processed to harden the soft endosperm to enable it to withstand the mechanical impact during

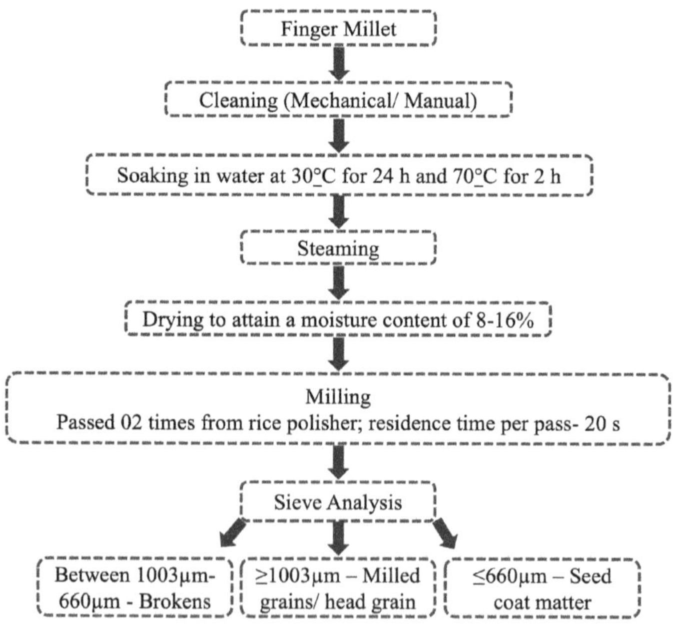

FIGURE 5.1
Flowchart for preparation of hydrothermally treated and milled finger millet.

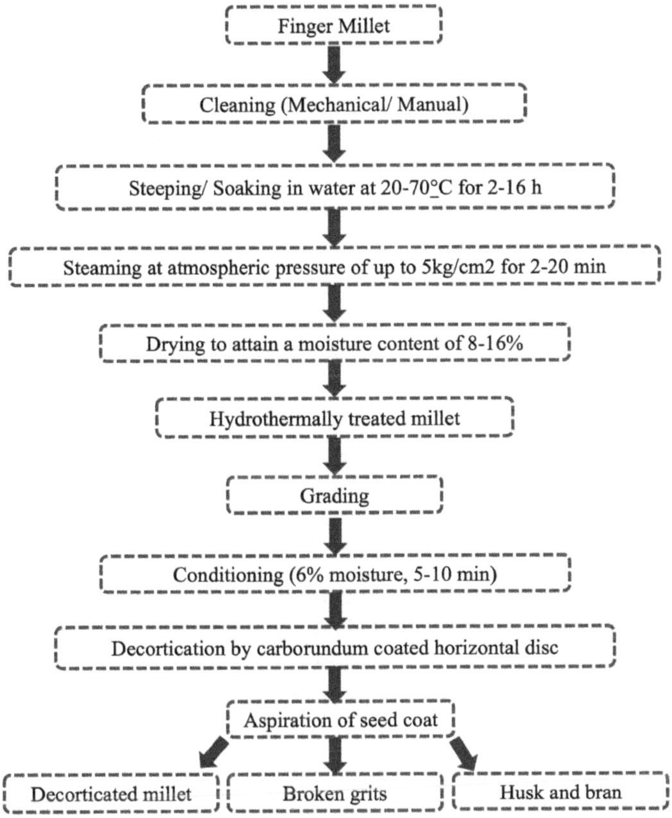

FIGURE 5.2
Flowchart of decortication process.

decortication Figure 5.2 (Hejazi & Orsat, 2016). The decorticated finger millet could be cooked as discrete grains similar to rice. However, the glycaemic property of decorticated finger millet (cooked in the discrete form similar to rice grains) is yet to be established (Shobana et al., 2009). In some cases, pearling or decortication is utilized to dehusk the finger millet grain; it brings about the crushing of both the seed coat and endosperm. Thus, finger millet is perpetually pounded alongside the seed to plan whole meal. Diffusive sheller can likewise be utilized to dehull/decorticate the little millets.

5.2.2.3.1 Process of Decortication

a. *Hydration:* The process typically begins with soaking the finger millet grains in potable water. This step is important as it prepares the grains for subsequent treatments by increasing their moisture content.
b. *Heat Treatment*: After soaking, the hydrated millet is subjected to heat treatment, which can involve steaming or boiling. This step induces starch gelatinization and softens the endosperm, making it easier to

separate the seed coat during milling. Optimal conditions for steaming include specific time and pressure settings, which have been shown to enhance the yield of decorticated grains.

c. *Drying:* Following heat treatment, the millet grains are dried to achieve a specific moisture content. This step is critical as it affects the hardness of the grains, which in turn influences the efficiency of the decortication process.
d. *Decortication:* The dried millet is then milled using appropriate machinery, such as emery-coated abrasive mills. During this stage, the seed coat is mechanically separated from the endosperm. The effectiveness of this step is influenced by the moisture content and the specifics of the milling process.
e. *Separation:* After milling, the decorticated millet is separated from the detached seed coats such as husk and bran, often using methods like winnowing or aspiration to ensure a clean product.

Importance of Decortication

- *Nutritional Quality*: Decorticated finger millet has improved nutritional quality, as the removal of the seed coat reduces the coarse fibrous texture and dark color associated with whole grain millet. This makes the flour more acceptable to consumers, especially those unfamiliar with millet.
- *Culinary Versatility*: The decorticated millet can be cooked similarly to rice or further processed into flour for various culinary applications, enhancing its usability in traditional and modern recipes.
- *Yield Optimization*: Research indicates that optimizing the conditions for decortication, including moisture content and steaming parameters, can significantly improve the yield of decorticated millet, which is vital for commercial processing.

The decortication process in milling finger millet is essential for enhancing its cooking quality, nutritional value, and consumer acceptance, while also optimizing processing efficiency (Smith & Johnson, 2018). The milling process also affects the nutritional profile of the flour. Research by Katti et al. (2008) showed that different milling techniques, such as roller mills and emery plate mills, impact the starch granules and nutrient content of the flour. Hydrothermal processing and decortication processes are known to influence the starch properties, protein content, and mineral availability in the flour.

5.3 Malting

Malting is a traditional process that involves soaking grains to initiate germination, followed by drying to stop the germination process. Germination protects the grains from fungal infection and has a dominant effect on numerous biochemical changes that enhance the nutritional value of malted products. Among the various tropical cereals, finger millet has good malting characteristics. Generally, barley is preferred among cereals for malting both in brewing and in food industries. However, attempts were made as early as 1939 to study the malting characteristics of finger millet, and Sastri (1939) reported the conditions for producing good-quality malt. The malt flour is a good source of amylases and is hence termed "Amylase-rich food" (Rajasekhar et al., 2018). Malt flour is a substitute for maltodextrin and can be blended with milk and spray-dried to prepare infant food (Malleshi & Desikachar, 1986a). During germination, amylases partially hydrolyse the starch to lower molecular weight carbohydrates such as oligo- and disaccharides, resulting in reduced water-holding capacity and increased energy density (Malleshi & Desikachar, 1982).

Malting of finger millet is regularly optimized for specialty food applications. During this process, the bioavailability of proteins, carbohydrates, and minerals improves. Some B-complex vitamins are synthesized, and anti-nutritional factors are reduced (Hejazi & Orsat, 2016; Rajasekhar et al., 2018). Malting involves soaking viable seeds in water to hydrate and facilitate sprouting Figure 5.3. These sprouts are then kiln-dried, and finally, the rootlets are

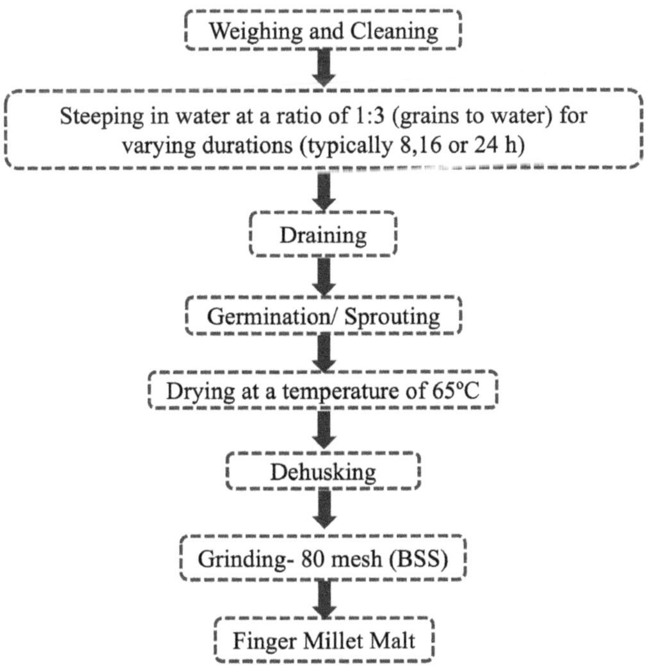

FIGURE 5.3
Flowchart of finger millet malting.

separated manually by rubbing. Each of these steps affects the quality of the malt (Hejazi & Orsat, 2016; Malleshi & Desikachar, 1982). Seed germination is the most crucial step as hydrolytic enzymes are developed, which transform the endosperm and increase nutritional properties (Hejazi & Orsat, 2016; Malleshi & Desikachar, 1986a). Malting of finger millet has been effectively used for producing various health foods such as infant food, weaning food, milk-based beverages, and confectionery products (Banusha & Vasantharuba, 2013). It leads to the expansion of α- and β-amylase during germination, which also contributes to the development of a desirable aroma during kilning and makes finger millet a superior grain for malt foods (Verma & Patel, 2013).

5.3.1 Steeping

Finger millet grains undergo cleaning to eliminate dirt and impurities. Steeping follows, where grains are submerged in water at a 1:3 grain-to-water ratio for 8–24 hours. To prevent fungal growth, the water is changed every 4 hours. A 0.2% sodium benzoate solution may be added to inhibit microbial activity. This process softens the grains and prepares them for malting (Banusha & Vasantharuba, 2013).

5.3.2 Germination

After steeping, the grains are drained and spread on muslin-lined trays in a controlled environment (25°C to 30°C). Sprinkling is done to maintain moisture. Germination lasts 24–72 hours, typically 48 hours, to maximize protein digestibility and reduce anti-nutritional factors (Hejazi & Orsat, 2016; Malleshi & Desikachar, 1982).

5.3.3 Drying

Following germination, grains are dried at approximately 65°C using an airflow drier. This halts germination, reduces moisture content, and prepares the grains for storage and processing (Hejazi & Orsat, 2016).

5.3.4 Dehusking

The dried grains are dehusked manually by rubbing to remove outer layers and rootlets. Dehusking improves texture, reduces fiber, and enhances palatability (Malleshi & Desikachar, 1982).

5.3.5 Grinding and Sieving

The malted grains are ground into fine flour using a grinder and sieved (typically 80 mesh) for consistency and suitable particle size for food applications (Hejazi & Orsat, 2016; Rajasekhar et al., 2018).

Malting increases hydrolytic enzymes. Varieties forming higher enzyme levels show low malting loss and are preferred for malting (Malleshi & Desikachar, 1982). Malting raises threonine, tryptophan, and protein content. Germination for 48 hours significantly improves in vitro extractability of calcium, zinc, and iron (Mamiro et al., 2001). Malting decreases dietary fiber and raises reducing sugar from 1.44% to 8.36% and total sugar from 1.5% to 16.0%. Malting also enhances protein content; malting finger millet at 48 hours/30°C increased protein from 6.42 g/100 g to 7.32 g/100 g (Hejazi & Orsat, 2016). Finger millet's superior malting properties make it ideal for malt-based products, enhancing its nutritional value and digestibility (Hejazi & Orsat, 2016; Malleshi & Desikachar, 1982; Verma & Patel, 2013).

5.4 Fermentation

Fermentation is a biochemical process in which carbohydrates are broken down into alcohol and acids by microorganisms under aerobic or anaerobic conditions. This process is widely utilized to produce a variety of fermented foods and beverages, enhancing their flavor, digestibility, and nutritional profile. In the case of finger millet, fermentation plays a crucial role in improving the sensory attributes and health benefits of the final products. The action of microorganisms during fermentation leads to desirable changes in flavor, digestibility, and the enhancement of minor nutrients and functional components, including antioxidants, phenolics, and phytochemicals (Antony et al., 2009; Banusha & Vasantharuba, 2013). Fermentation is particularly significant in the production of traditional foods such as *Idli, Dosa,* and *Bhathura,* which are staples in Indian cuisine. Globally, cereals including finger millet are increasingly used in the brewing industry for alcoholic beverages. Prior to brewing, processes such as germination and malting are employed (Chandrasekhara & Swaminathan, 1954). For finger millet, this involves steeping the grains for approximately 16 hours, followed by germination for 2–3 days, drying, and grinding into malted flour for beer production (Mugocha et al., 2000).

Different fermentation techniques, characterized by variations in starter cultures, fermentation periods, and temperatures, have been explored to optimize the process (Figure 5.4). Controlled fermentation, achieved through enzymatic treatment followed by fermentation, has shown a reduction in fermentation time from 48 hours in natural conditions to just 12 hours, without adversely affecting pH, phytate, tannins, or mineral extractability (Antony et al., 2016).

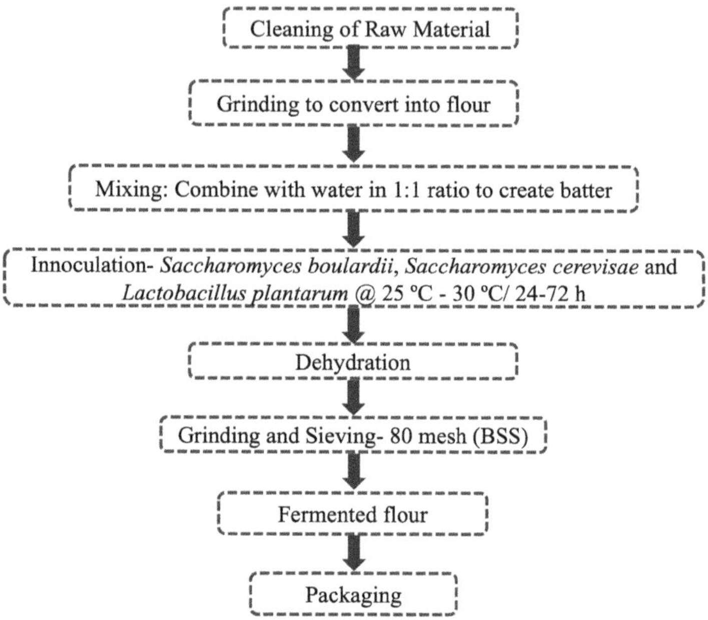

FIGURE 5.4
Process flowchart of fermented flour.

5.4.1 Cleaning and Grinding

Finger millet grains are thoroughly cleaned to remove impurities, ensuring a high-quality starting material. The cleaned grains are ground into flour using a mixer grinder, yielding a fine consistency suitable for fermentation.

5.4.2 Mixing and Inoculation

The flour is mixed with water in a 1:1 ratio to form a batter. Selected starter cultures such as *Saccharomyces boulardii*, *Saccharomyces cerevisiae*, and *Lactobacillus plantarum* are added to initiate fermentation. The batter is fermented at 25°C to 30°C for 24–72 hours, depending on the starter used.

5.4.3 Drying

Post-fermentation, the batter is dried using oven or spray drying methods. Oven drying applies controlled heat for moisture removal, while spray drying atomizes the batter into hot air, forming dry particles. This dried flour has a longer shelf life and improved texture and flavor, making it suitable for varied applications.

5.4.4 Milling and Sieving

The dried fermented flour is ground again and sieved to achieve a uniform particle size (~80 mesh, BSS), ensuring consistency for subsequent processing. Fermentation has been demonstrated to significantly reduce anti-nutrients such as tannins, oxalates, and phytates, which otherwise hinder mineral absorption. Studies have shown that fermentation lowers phytate levels, improving mineral bioavailability, particularly calcium, iron, and zinc (Mamiro et al., 2001). Fermentation can also raise protein content, with increases of up to 42% observed after 72 hours of fermentation (Antony et al., 1996; Banusha & Vasantharuba, 2013). Additionally, fermented millet samples display improved carbohydrate profiles, with reductions in total carbohydrates and increases in free sugars that enhance sweetness and palatability. For instance, after 96 hours of fermentation, an increase in free sugars and a corresponding reduction in starch was recorded (Antony et al., 1996). Fermentation also boosts phenolic content and antioxidant activity (Chethan & Malleshi, 2007), making fermented millet nutritionally superior. The optimal fermentation duration appears to be 48–72 hours at around 28°C ± 2°C to maximize nutritional benefits (Antony et al., 1996; Mamiro et al., 2001).

5.5 Puffing

Puffing or popping is a traditional processing technique that increases the convenience and sensory appeal of millet-based foods. The application of high heat causes the internal moisture to vaporize rapidly, expanding the grain structure and resulting in a light, crispy texture (Figure 5.5) (Malleshi & Desikachar, 1986a).

5.5.1 Cleaning and Grading

Finger millet grains are cleaned and graded for size uniformity, which is crucial for consistent puffing.

5.5.2 Soaking

Grains are soaked at 28°C–30°C for 8–30 hours to absorb moisture and prepare for germination.

5.5.3 Germination

After soaking, grains are wrapped in muslin and incubated to germinate for 24–48 hours. Germination activates enzymes that break down complex carbohydrates (Shobana et al., 2009).

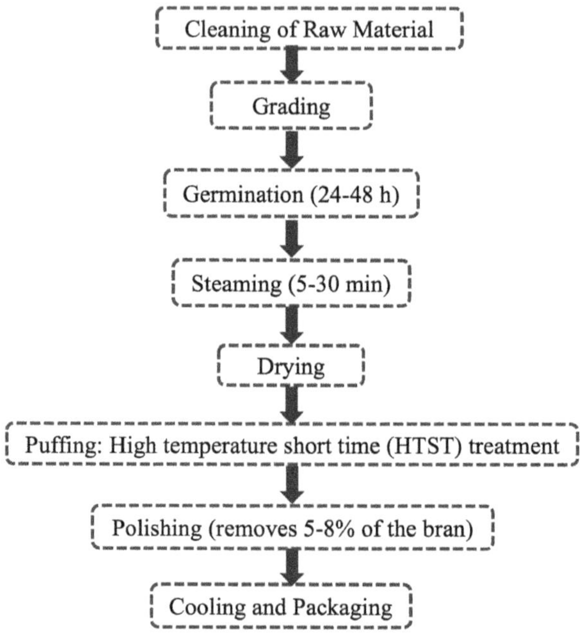

FIGURE 5.5
Process flowchart of puffing process.

5.5.4 Steaming

Steaming for 5–30 minutes gelatinizes the starch, facilitating puffing.

5.5.5 Drying

Grains are dried to 8–16% moisture using sunlight or dryers to prevent spoilage and ensure readiness for puffing.

5.5.6 Puffing

High-temperature short-time (HTST) treatment using hot air, sand, salt, or gun puffing is applied. This causes water inside the grains to vaporize, expanding the structure.

5.5.7 Polishing

Polishing removes 5–8% of the bran using an emery polisher, refining the product texture (Malleshi & Desikachar, 1982).

5.5.8 Cooling and Packaging

Grains are cooled and packaged to preserve their crispy texture.Puffing impacts the nutritional profile, typically reducing fat and increasing mineral bioavailability, particularly iron. It is a rapid gelatinization process that enhances starch digestibility and flavor. When blended with malted or roasted legumes, puffed flour is used in weaning foods and RTE snacks. RTE- Ready-to-Eat (RTE) finger millet products are food items prepared from decorticated or malted finger millet grains, processed and packaged to be consumed directly without further cooking. These products have gained prominence due to the increasing consumer demand for convenient, nutritious, and functional foods. Common processing techniques for RTE finger millet products include flaking, extrusion, puffing, and roasting, often combined with thermal treatments such as baking or steaming to ensure microbial safety and improve shelf life (Malleshi & Desikachar, 1986a). Temperature and moisture control are important as optimal puffing occurs between 100°C and 200°C with 8–16% moisture content.

5.6 Extrusion

Extrusion is a HTST processing technique that transforms raw materials into ready-to-eat food products through a combination of mechanical and thermal means. This innovative method is widely utilized in the production of various food items, including snacks and cereals, by gelatinizing and cooking the ingredients until they are fully processed. The extrusion process employs heat, steam, and pressure to create a diverse range of food products. Finger millet has emerged as a promising cereal for various food applications, including the production of vermicelli. A study by Verma and Patel (2013) demonstrated that incorporating finger millet flour of different particle sizes into wheat extruded vermicelli resulted in higher sensory scores when medium particle size flour (50%) was used in hot water dough preparation followed by extrusion. While extruded products made from finger millet may exhibit changes in color, texture, and storable properties, they also offer advantages such as increased dietary fiber content and cost-effectiveness (Singh et al., 2012).

5.6.1 Mechanism of Extrusion

During the extrusion process, grains, such as finger millet, are subjected to high pressure as they move through an extruder. The conditions within the extruder cause the deactivation of anti-nutritional factors present in the grains due to the high temperatures achieved during processing (Hejazi & Orsat, 2016). This thermal treatment not only enhances the safety and digestibility of the product but also alters the protein content. Proteins undergo denaturation, a process that changes their structure and solubility, particularly at temperatures exceeding 50°C. Research indicates that the average denaturation temperature of proteins in Proso millet is approximately 82.1°C ± 3.5°C (Smith & Johnson, 2018).

As a result of the extrusion process, certain hydrophobic properties of the proteins may decrease; however, emulsion stability remains largely unaffected (Hejazi & Orsat, 2016). Emulsion stability is crucial as it influences the color, texture, and overall succulence of the millet products. Maintaining emulsion stability prevents the separation of oil and water components, ensuring a desirable blending effect that preserves the quality, freshness, and integrity of the final cooked products. Extrusion not only improves the feed value of the processed grains by inactivating anti-nutritional factors but also promotes the availability of undegraded digestible dietary protein, particularly beneficial for ruminant animals (Verma & Patel, 2013). Advances in technology have facilitated the development of extrusion cooking, which can produce low-energy foods, including those with reduced fat and protein content (Singh et al., 2012). This technology is particularly valuable for creating ready-to-eat snacks that offer health benefits.

Extrusion Process Steps:

i. *Preparation of Raw Material*: The first step involves grinding finger millet grains into flour, typically using a hammer mill (Katti et al., 2008). This flour is then conditioned to achieve specific moisture content levels, which can range from approximately 14.95% to 25.05% (wet basis), depending on the desired characteristics of the final product (Rajasekhar et al., 2018).

ii. *Extrusion Conditions:* The extrusion process requires heating the flour in a barrel at temperatures typically between 103°C and 136.82°C, with optimal temperatures around 130°C to maximize expansion and functional properties (Verma & Patel, 2013). Achieving the right feed moisture content is crucial; an optimal level of around 21.3% has been identified for achieving desirable expansion ratios and functional properties. The screw speed during extrusion can vary, with effective speeds around 375 rpm. However, studies suggest that screw speed has a lesser impact on the final product compared to moisture content and temperature (Verma & Patel, 2013).

iii. *Extrusion Process*: The conditioned flour is fed into a twin-screw extruder, where it is subjected to high pressure and shear. This combination of conditions facilitates starch gelatinization and the development of the extrudate's texture. The extrudate is then cut into desired shapes using a die face cutter (Verma & Patel, 2013).

iv. *Cooling and Storage:* After extrusion, the products are cooled to room temperature and stored in sealed polyethylene bags until further analysis or packaging (Verma & Patel, 2013).

5.6.2 Applications of Extruded Finger Millet

Extruded finger millet products can be utilized in various food applications, including:

Snack Foods: The extrusion process is commonly employed to create expanded snack products that appeal to consumers (Singh et al., 2012).

Infant Foods: Due to its nutritional benefits, extruded finger millet is suitable for formulating healthy infant foods (Malleshi & Desikachar, 1986a).

Composite Foods: Finger millet can be blended with other flours (e.g., rice, corn) to enhance the nutritional profile of various food products (Verma & Patel, 2013).

5.6.3 Challenges in Extruding Finger Millet

Despite its advantages, the extrusion of finger millet presents several challenges that can impact the quality and efficiency of the process:

Moisture Control: Maintaining optimal moisture content is critical for successful extrusion. Variations in moisture can lead to inconsistent extrudate properties, such as expansion ratio and texture. High moisture levels may increase the WAI but could decrease the WSI, negatively affecting the overall quality of the extrudate (Rajasekhar et al., 2018).

Temperature Management: Precise temperature control is essential to ensure proper gelatinization of starches. Inadequate temperature management can result in undercooked or overcooked products, adversely affecting texture and nutritional quality (Hejazi & Orsat, 2016; Verma & Patel, 2013).

Ingredient Variability: Finger millet exhibits variability in its physical and chemical properties due to differences in grain size, shape, and composition among varieties (Katti et al., 2008). This variability can complicate the extrusion process, necessitating adjustments to achieve consistent product quality. The interaction between finger millet flour and other ingredients, such as starches or proteins, must also be carefully managed for optimal blending and extrusion outcomes (Verma & Patel, 2013).

Shelf Life and Quality Retention: Extruded finger millet products often face challenges related to shelf life due to intrinsic enzyme activity, which can lead to rancidity and loss of quality over time. Ensuring the stability of the extrudates during storage requires effective pretreatments and optimal packaging solutions to minimize moisture absorption and oxidation (Verma & Patel, 2013).

Market Acceptance and Awareness: Despite the nutritional benefits of finger millet, there is limited awareness and acceptance among consumers. This poses a challenge for manufacturers in promoting extruded products made from finger millet, as market dynamics can affect the viability and profitability of these products (Singh et al., 2012).

Complexity of the Extrusion Process: The extrusion process is inherently complex, involving multiple variables (such as feed rate, screw speed, and blending ratios) that must be optimized to achieve desired product characteristics. This complexity can lead to difficulties in scaling up production and ensuring consistent quality across batches (Verma & Patel, 2013).

In summary, addressing the challenges associated with the extrusion of finger millet requires careful optimization of processing parameters, a thorough understanding of the raw materials, and effective marketing strategies to enhance consumer acceptance of finger millet-based products in the market. By overcoming these challenges, manufacturers can leverage the benefits of extrusion technology to produce high-quality, nutritious, and appealing food products.

5.7 Popping

This is one of the important processing techniques widely used to prepare ready-to-eat products, which involves HTST treatment to finger millet using sand as heat transfer media, where finger millet starch gelatinizes and the endosperm bursts open. Popped finger millet possesses a highly desirable flavor and aroma. It is used as a snack after seasoning with spices and condiments. It is a whole-grain product rich in macronutrients, micronutrients, and dietary fiber, and is usually mixed with vegetable or milk protein sources such as popped Bengal gram, milk powder, and oil seeds, sweetened with jaggery or sugar to prepare ready-to-eat nutritious supplementary food, as noted by Katti et al. (2008). In recent days, finger millet is puffed using an air popper device, which is considered superior in terms of quality, with no sand and low salt in the product compared to the traditional method using sand or salt as heat transfer media. Recently, expanded finger millet has also been developed from decorticated finger millet. The dietary fiber content of the expanded finger millet is lower than that of popped finger millet as it is prepared from decorticated finger millet, which is devoid of the seed coat. Fermented beverages from finger millet are also common.

5.8 Baking

Baking is a traditional method used to produce various bakery products. For finger millet, baking involves blending the flour with other flours, such as wheat or maida, to produce baked goods. Incorporating finger millet flour into baked products often requires adjustments due to its different starch properties compared to wheat flour. Simple addition can result in undesirable textures and flavors, so modification through processes like puffing or roasting can improve the characteristics of the baked products. Finger millet flour is used in baking to create products like bread, cakes, and cookies. Research has shown that blending finger millet flour with wheat flour can improve the texture and quality of baked goods, though further optimization is needed to achieve desired results (Singh et al., 2012; Verma & Patel, 2013).

5.9 Combination Processing

Combination processing involves using multiple processing methods in sequence to achieve synergistic effects. This approach can enhance the nutritional quality, flavor, and functional properties of finger millet products. Combining processes such as germination, malting, fermentation and puffing can improve the overall quality of the final products. For instance, combining germination with fermentation can enhance the bioavailability of nutrients like iron and zinc (Antony et al., 1996; Mamiro et al., 2001). Combination processing is used to create various weaning foods, supplementary foods, and health products. Studies have shown that combining different processing methods can lead to improved protein quality, mineral extractability, and reduced anti-nutrient factors (Chethan & Malleshi, 2007; Malleshi & Desikachar, 1982).

5.10 Challenges in the Processing of Finger Millet

5.10.1 Agronomic Challenges

i. *Inconsistent Rainfall and Crop Management*: Finger millet is typically cultivated in areas with marginal soils and variable rainfall. Fluctuating weather patterns and inadequate crop management can result in lower yields and compromised grain quality.

ii. *High Input Costs*: The expenses associated with fertilizers and pesticides are significant, creating barriers for smallholder farmers who predominantly grow finger millet. This economic burden can deter optimal farming practices, leading to diminished production levels (Singh et al., 2012).

5.10.2 Processing Limitations

i. *Outdated Processing Technologies*: Traditional processing methods are prevalent in rural regions, often being labour-intensive and time-consuming. There is a pressing need for modern processing technologies that can enhance efficiency and improve product quality. The lack of such innovations restricts the production of value-added products that could attract urban consumers (Premavalli & Madhura, 2012).

ii. *Presence of Anti-Nutritional Factors*: Finger millet contains anti-nutritional components, such as tannins, phytic acid, and trypsin inhibitors, which can impede nutrient absorption. Although techniques like soaking, germination, and fermentation can mitigate these factors, limited awareness and access to these methods hinder their widespread application (Chethan & Malleshi, 2007; Shobana, 2009).

5.10.2.1 Market and Consumption Issues

i. *Decreasing Demand*: Consumer preferences have shifted towards more refined cereals like rice and wheat, often due to convenience and taste. This trend has resulted in a decline in market demand for finger millet, especially in urban areas where it is less recognized (Verma & Patel, 2013).

ii. *Limited Product Availability*: There is a scarcity of finger millet products tailored to urban tastes. This gap in the market restricts broader acceptance and consumption of finger millet, despite its nutritional advantages (Smith & Johnson, 2018).

5.10.3 Nutritional Awareness

Underappreciation of nutritional benefits: Despite being rich in essential nutrients such as calcium and dietary fiber, public awareness regarding the health benefits of finger millet is insufficient. Educational initiatives are necessary to promote its consumption and emphasize its advantages over more commonly consumed grains (Shobana et al., 2009).

5.10.4 Physical Characteristics

a. *Color*: Finger millet has a natural reddish-brown to dark brown hue due to pigments in the bran layer. This coloration can be unappealing for certain products, necessitating decortication or other processing methods to achieve a lighter flour.
b. *Small Grain Size*: The small size of finger millet grains complicates processing. It can lead to higher losses during milling and other processing methods, as the tiny grains are more prone to breakage and dust.
c. *Intact Endosperm*: The endosperm of finger millet is tightly bound to the bran layer, making it challenging to mill into refined flour. Hydrothermal processing and decortication can help separate the endosperm from the bran (Rajasekhar et al., 2018).
d. *Fat Composition*: Finger millet contains a higher proportion of unsaturated fatty acids compared to other millets, making the flour more susceptible to rancidity and reducing its shelf life. Proper storage conditions and packaging are essential to prevent lipid oxidation.
e. *Nutritional Compounds*: Finger millet contains anti-nutrients like phytates and tannins that can diminish nutrient bioavailability. Processing methods such as malting, fermentation, and extrusion can help lower these anti-nutrient levels (Antony et al., 1996; Malleshi & Desikachar, 1986a).

5.11 Conclusion

Processing of finger millet (*Eleusine coracana*) plays a pivotal role in enhancing its nutritional quality, functional attributes, and consumer acceptability. Primary processing methods such as decortication and milling facilitate the removal of outer layers, thereby improving palatability and reducing anti-nutritional factors. Secondary processing techniques—including malting, fermentation, puffing, extrusion cooking, and baking—have demonstrated significant influence on the grain's physicochemical, nutritional, and sensory properties. Optimization of these techniques is essential to improve the bioavailability of micronutrients such as iron, calcium, and zinc, which are often bound by anti-nutritional compounds like phytates, tannins, and oxalates. Soaking, a widely adopted pre-treatment, has been shown to reduce phytate content substantially, thus enhancing mineral bioaccessibility. Similarly, integrated processing methods, such as combining germination with fermentation or malting, result in synergistic improvements in protein digestibility, enzymatic activity, and reduction of anti-nutrients.

Moreover, the development of diversified value-added products, such as malted flours, extruded snacks, baked goods, and fermented beverages, offers potential for increasing the marketability and consumption of finger millet, particularly in urban settings. Future advancements should focus on scaling up efficient and economically viable processing technologies, coupled with mechanization and standardization, to ensure quality consistency and nutritional retention. Overall, comprehensive application and refinement of processing interventions are critical for enhancing the functional utility and dietary relevance of finger millet in both traditional and contemporary food systems.

References

Abulude, F. O., Lawal, L. O., & Kayode, A. O. (2005). Effect of processing on some functional properties of millet (*Eleusine coracana*) flour. Journal of Food Technology, 3(3), 460–463.

Antony, A., Sripriya, G., & Chandra, T. S. (1996). Effect of fermentation on the primary nutrients in finger millet (*Eleusine coracana*). The Journal of Agricultural and Food Chemistry, 44, 2616–2618.

Banusha, S., & Vasantharuba, S. (2013). Effect of malting on nutritional contents of finger millet and mung bean. American-Eurasian Journal of Agriculture and Environmental Science, 13(12), 1642–1646.

Chandrasekhara, M. R., & Swaminathan, M. (1954). Enzymes of ragi (*Eleusine coracana*) and ragi malt. III. Pyro- and glycero-phosphatases. *Journal of Scientific and Industrial Research*, 13B, 492–496.

Chethan, S., & Malleshi, N. G. (2007). Finger millet polyphenols: Characterization and their nutraceutical potential. American Journal of Food Technology, 2, 582–592.

Hejazi, S. N., & Orsat, V. (2016) Malting process optimization for protein digestibility enhancement in finger millet grain. Journal of Food Science and Technology, 53, 1929–1938.

Katti, S., Kumar, S., & Malleshi, N. (2008). Studies on the effect of milling finger millet in different pulverisers on physicochemical properties of the flour. Journal of Food Science and Technology, 45, 398–405.

Kurien, P. P., & Desikachar, H. S. R. (1966). Preparation of refined white flour from ragi using a laboratory mill. Journal of Food Science and Technology, 3, 56–58.

Kurien, P. P., Joseph, K., Swaminathan, M., & Subrahmanyan, V. (1959). The distribution of nitrogen, calcium and phosphorus between the husk and endosperm of ragi (*Eleusine coracana*). Food Science, 8(10), 353–355.

Malleshi, N. G., & Desikachar, H. S. R. (1981a). Studies on the suitability of roller flour mill, hammer mill and plate grinder for obtaining a refined flour from malted ragi (*Eleusine coracana*). The Journal of Food Science and Technology, 18, 37–39.

Malleshi, N. G., & Desikachar, H. S. R. (1982). Formulation of a weaning food with low hot paste viscosity based on malted ragi (*Eleusine coracana*) and green gram (*Phaseolus radiatus*). The Journal of Food Science and Technology, 19, 193–197.

Malleshi, N. G., & Desikachar, H. S. R. (1986a). Studies on comparative malting characteristics of some tropical cereals and millets. Journal of the Institute of Brewing, 92, 174–176.

Mamiro, P., Van, J., Mbithi-Mwikya, S., & Huyghebaert, A. (2001). In vitro extractability of calcium, iron, and zinc in finger millet and kidney beans during processing. Journal of Food Science, 66, 1271–1275. https://doi.org/10.1111/j.1365-2621.2001.tb15200.x

Mugocha, P. T., Taylor, J. R. N., & Bester, B. H. (2000). Fermentation of a composite finger millet-dairy beverage. World Journal of Microbiology and Biotechnology, 16, 341–344.

Rajasekhar, M., Edukondalu, L., Smith, D. D., & Veeraprasad, G. (2018). Effect of hydrothermal treatment on milling characteristics of finger millet (*Eleusine coracana*). *International Journal of Current Microbiology and Applied Sciences*, 7(10), 1804–1811. https://doi.org/10.20546/ijcmas.2018.710.206.

Sastri, B. N. (1939). Ragi, *Eleusine coracana* Gaertn—A new raw material for the malting industry. Current Science, 1, 34–35.

Shobana, S. (2009). Investigations on the carbohydrate digestibility of finger millet with special reference to the influence of its seed coat constituents. Ph.D. Thesis. University of Mysore, Mysore.

Shobana, S., Sreerama, Y. N., & Malleshi, N. G. (2009). Composition and enzyme inhibitory properties of finger millet (*Eleusine coracana* L.) Seed coat phenolics: Mode of inhibition of α-glucosidase and α-amylase. Food Chemistry, 115, 1268–1273.

Singh, P., & Raghuvanshi, R. S. (2012). Finger millet for food and nutritional security. African Journal of Food Science, 6(4), 77–84. https://doi.org/10.5897/AJFSX10.010

Smith, J. A. (2022). Agricultural value addition: Transforming raw commodities for increased profitability. Journal of Agricultural Economics, 36(2), 123–145. DOI: 10.1111/jagec.12345

Verma, V., and S. Patel. "VALUE ADDED PRODUCTS FROM NUTRI-CEREALS: FINGER MILLET (ELEUSINE CORACANA)". Emirates Journal of Food and Agriculture, vol. 25, no. 3, Nov. 2017, p. 169/176, doi:10.9755/ejfa.v25i3.10764.

6

Physico-Chemical Properties, Modifications, and Applications of Finger Millet Starch

Milad Tavassoli, Reza Abedi-Firoozjah, and Behnam Bahramian

6.1 Introduction

The term "millet" describes the small-grained grasses of the Poaceae family, namely the Paniceae and Chlorideae tribes (Fig. 6.1). This plant has been a staple of humans for about 10,000 years, before the cultivation of rice or wheat grains (Lu *et al.*, 2009). The most commonly grown types of millet in recent times have been pearl millet (*Pennisetum glaucum*; synonyms: *P. typhoides, P. typhoideum, P. americanum*), proso (*Panicum miliaceum*), and foxtail (*Setaria italica*) millet. Browntop millet (*Brachiaria ramosum*), white fonio (*Digitaria exilis*), finger millet (*Eleusine coracana*), black fonio (*Digitaria iburua*), shama millet (*Echinocloa colona*), and less frequently grown millet species are still remarkably productive locally (Chandi & Annor, 2015). Millets are highly productive on marginal soils, have a short growth cycle, and can thrive in various climatic conditions, including heat and drought. Due to their resilience, they are well-suited for enhancing food availability in semi-arid parts of Asia and Africa, where the production of other significant crops often proves unsuccessful (Chandi & Annor, 2015). In 2012, production of all millets worldwide was approximately 704,239 tons. The largest producer is India, accounting for around 273,000 tons, with Niger coming in second, which produced almost 106,427 tons above its needs. Niger followed it with surplus production of 106,427 tons and Nigeria with 60,800 tons (Canton, 2021).

Millet is a highly nutritious grain that offers a variety of health benefits, making it an exceptional addition to many diets. Rich in essential nutrients, millet contains significant amounts of calcium, which is crucial for maintaining strong bones and teeth. It is also a valuable source of B vitamins, including niacin, riboflavin, and folic acid, which play vital roles in energy metabolism and overall brain health. Additionally, millet is abundant in iron, an essential mineral that supports red blood cell production and helps prevent anemia. Moreover, millet is a source of polyphenols, which are antioxidants that can

DOI: 10.1201/9781003535119-6

FIGURE 6.1
(a) Plant, (b) leaves, (c) seeds, and (d) flour of finger millet.

help neutralize free radicals in the body. This property may contribute to reducing oxidative stress and the risk of chronic diseases. The phosphorus found in millet, primarily in the form of phytate, is important for bone health and contributes to the body's energy production. Dietary fiber is another key component of millet, promoting healthy digestion and aiding in weight management by providing a feeling of fullness. The presence of healthy lipids in millet further supports heart health by helping to manage cholesterol levels (Fig. 6.2) (Shahidi & Chandrasekara, 2013). One of the remarkable qualities of millet is its gluten-free status, making it an excellent alternative for individuals with celiac disease or gluten sensitivities. This inclusivity makes it a viable option for maintaining a balanced diet without the adverse effects associated with gluten consumption (Taylor & Emmambux, 2008). Moreover, research has shown that millet possesses hypoglycemic properties, which means it may assist in lowering blood sugar levels. However, further studies are needed to fully understand the mechanisms behind this phenomenon

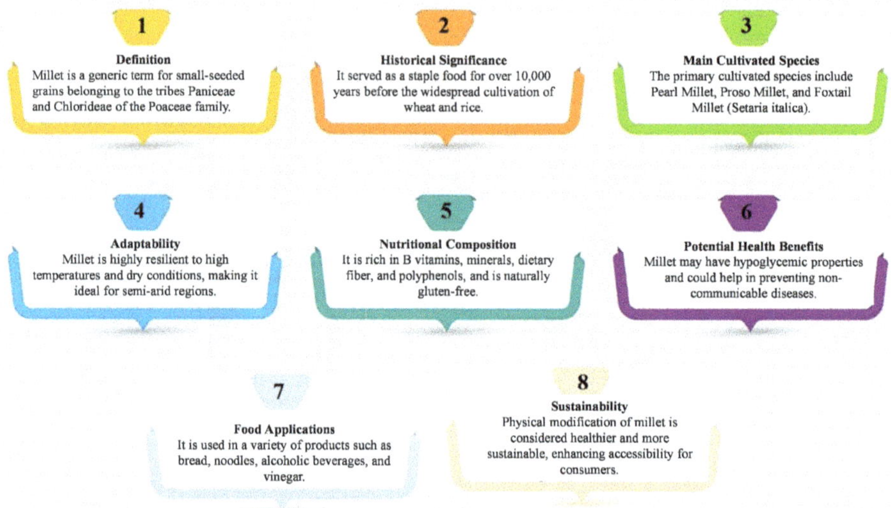

FIGURE 6.2
Summary of the key aspects of millet and their relevance in food production and health.

and how millet can be effectively utilized for managing diabetes (Annor *et al.*, 2013). Collectively, these nutritional attributes and bioactive compounds position millet as a crucial functional food that has the potential to prevent or delay the onset of non-communicable diseases, such as cardiovascular diseases (CVDs) and type 2 diabetes. Incorporating millet into a regular diet can therefore contribute significantly to overall health and wellness (Saleh *et al.*, 2013; Shahidi & Chandrasekara, 2013).

Finger millet has 15–20% dietary fiber, 65–75% carbs, and 5–10% protein. Millet is a resilient crop in certain regions and is the sixth most-produced grain globally (Saini *et al.*, 2021). Starch is a renewable bio-macromolecule found in different plants and crops and it is utilized in various large-scale uses and industrial applications. Starch, a natural condensation polymer, is essential in the human diet and can be readily chemically modified because of its hydroxyl groups. This makes starches highly versatile for food applications, given their low cost, widespread availability, and the ease with which their physico-chemical properties can be altered through various modification techniques (Ashogbon, 2021; Bangar *et al.*, 2024). Because of its poor gel stability, susceptibility to heat and shear, and low solubility in cold water, natural millet starch has limited use in both culinary and industrial settings (Li *et al.*, 2019; Thangaraju *et al.*, 2022). Starches are frequently present in assorted food items, such as soups, stews, gravies, pie fillings, sauces, and custards. They function as thickening and gelling agents, bulking agents, and adhesives, significantly enhancing the texture of meals.

This chapter provides a thorough overview of the current knowledge regarding the extraction, chemical composition, physical and chemical properties, susceptibility to enzymes, modifications, and applications of millet starch. It also explores the morphological, functional, and technological attributes, various modification techniques, and uses of finger millet starch (FMS). The chapter emphasizes the importance of additional research to improve the quality and expand the utilization of this particular type of starch.

6.2 An Overview of Finger Millet Starch

6.2.1 The Health Advantages of Finger Millet

The seed coat of finger millet, frequently called ragi, consists of five layers that are abundant in dietary fiber and functional antioxidant agents. In terms of nutritional composition, finger millet is on par with other millets and cereals, containing approximately 81.5% carbohydrates, 65–75% starch, 9.8% protein, 18–20% dietary fiber, 1–1.7% fat, 2.7% minerals, and 4.3% crude fiber (Saleh *et al.*, 2013). Finger millet is an impeccable source that is associated with various health benefits (Fig. 6.3). High fiber content and an amino

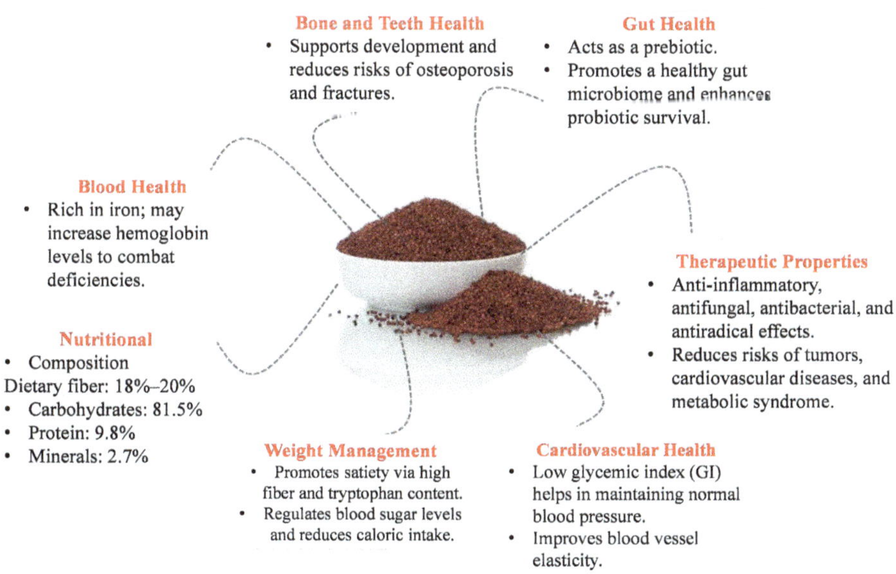

FIGURE 6.3
Health benefits of finger millet.

acid called tryptophan promote satiety, which by slowing down digestion and absorption, helps in maintaining weight. This further regulates blood sugar levels and can eventually result in the intake of fewer calories, hence leading to healthy weight loss (Geetha *et al.*, 2020). In addition, finger millet, containing 344 mg of calcium in its serving, contributes a great deal to good bone and teeth development in both children and older people, saving them from osteoporosis and fractures. Its low Gastric Intestinal (GI), in combination with a high level of fiber, It is highly beneficial for regulating blood pressure and promoting the health of blood vessels by raising their elasticity and promoting the relaxing of vessels (Anitha *et al.*, 2021). Finger millet contains significant amounts of iron and calcium and can increase hemoglobin levels, thus fighting hemoglobin and iron deficit. More so, finger millet acts in the role of a prebiotic and promotes robust intestinal microbiota by increasing the effectiveness and survival of probiotics in the digestive system. It is particularly beneficial for pregnant women and people with diabetes (Dhanushkodi *et al.*, 2023). Finger millet possesses some unique therapeutic values owing to the presence of anti-inflammatory, antiradical, antifungal, and antibacterial properties. It reduces health risks related to tumors, CVDs, and metabolic syndrome by reducing blood sugar and lipid levels (Vedamanickam *et al.*, 2020). Further, the high-level content of minerals like magnesium and phosphorus in finger millet maintains bone health, energy production, and heart and brain functions, maintains kidney health, monitors blood sugar levels, and relaxes the body while maintaining gums and teeth health (Maharajan *et al.*, 2022).

6.2.2 Techniques for Obtaining Starch from Finger Millets

Granular starch is present and is extractable using various methods, including alkaline extraction, acid extraction, and the distilled water approach, incorporating techniques such as filtration, gravity sedimentation, and centrifugation (Fig. 6.4). To help liberate starch from the grain structure, immersing millet flour in distilled water for a duration of 12–24 h (temperature; 20°C–25°C). while being constantly stirred. This method emphasizes a simple soaking process for effective starch extraction. In contrast, acid treatment covers the reaction of millet flour with 0.1 M hydrochloric acid (HCl) in a 1:6 ratio, while extraction of starch is conducted within the range of 30°C–40°C for 8 h (El Halal *et al.*, 2019). This approach takes advantage of the prevalence of acid under such conditions to ensure better separation of starch over a while. On the other hand, in an alkaline method, polished millet is soaked in water and then ground in a wet grinder with more water (Gautam *et al.*, 2021). Here, the final mixture is passed through a sieve and the residue is ground and sieved repeatedly until all the starch is removed. It is then mixed with water and purified through various washing steps. First, it is treated with a 0.2 N NaOH solution to separate proteins from the starch. Finally, to maintain the pH at 6.5–7, a NaCl solution is used. Centrifugation

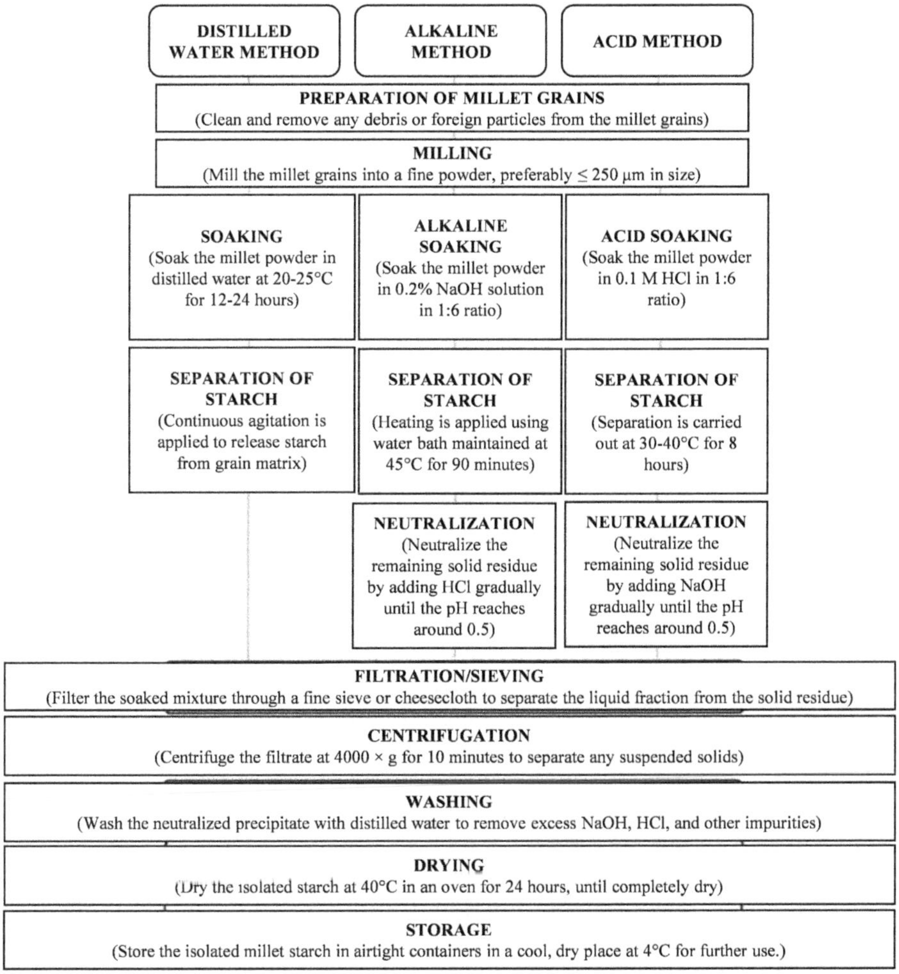

FIGURE 6.4
General procedures of starch isolation (Bhasin *et al.*, 2024).

is then carried out with the sediment. Toluene acts to denature the proteins in the final sediment, after which the upper layer that contains proteins is carefully discarded. This process of centrifugation and removal of proteins is generally done more than once to obtain highly purified starch (Amarnath *et al.*, 2023). After the starch is cleaned, it is dried, pulverized using a mortar and pestle, and then filtered via a mesh screen to ensure the same particle dimensions.

Ultimately, it is enclosed in polyethylene bags (Balakumaran *et al.*, 2023). Besides the ease with which the grinding process is facilitated, hydration during processing also increases access to nutrients and reduces anti-nutrients. However, prolonged hydration increases the avenues through which immune contamination can occur. A non-thermal technology still in investigation for the effective extraction of FMS is ultrasound-assisted hydration. Of the two processes, the alkaline process is preferable as it gives higher yields than using distilled water. Residual acid-isolated starch may exist acids and hence is not safe for human consumption (Yadav *et al.*, 2021).

6.2.3 Chemical and Structural Composition of Finger Millet Starch

Millets typically have a starch content ranging from 51% to 79%, comparable to other cereals. In finger millets specifically, starch levels range from 71% to 81% (Karki & Kharel, 2012). The starch in millet is made up of approximately 20–30% amylose and 70–80% amylopectin. All the other minor components may represent impurities in FMS and always contribute significantly to the functional properties of starch, as represented in Table 6.1 (Alcázar-Alay & Meireles, 2015). Millet starches predominantly consist of polar phospholipids, which consist of 89% of the total lipid composition. The rest of the composition is made up of nonpolar lipids, mainly in the form of triglycerides (Cornejo-Ramírez *et al.*, 2018). These lipids form a compound with the starch's amylose portion, which may limit the starch's ability to grow and flow, which might be attributed to hydrophobic bonds/cohesive properties of starches (Tester *et al.*, 2004). Proteins contain both hydrophilic and hydrophobic bonds, which affect their ability to interact with water and oil (Chandra *et al.*, 2015).

Dietary fiber have been shown to reduce the absorption of oil, therefore may reduce the ability to bind oil. Amylose is an important element in starch functionalities (D. N. Yadav & Rajan, 2012). As opposed to proso and foxtail starches, the starches from pearl and finger millet have longer amylose chains. Their extended glucan chains are situated between their branching points rather than being at the ends (Annor *et al.*, 2014a). When exposed to an iodine solution, the components of starch, specifically amylopectin and

TABLE 6.1

Starch Yield and Chemical Composition of Finger Millet Starches

Starch yield (%)	Amylose (%)	Lipid	Protein	Ash	References
63.4	32.4	–	–	–	Annor *et al.* (2014a)
60.37	–	–	2.71	2.70	Balasubramanian *et al.* (2011)
52.4	21.51	0.24	1.45	0.32	Lawal (2009)
–	28.40		0.58	1.60	Wu *et al.* (2014)

amylose, create complexes that impart a red color to amylose and a blue color to amylopectin. The branching nature imparts steric barrier and determines the capability of amylose and amylopectin to form helices with iodine molecules, as suggested from their blue intensity and selectivity for binding iodine (Zhu, 2014b).

The chain length (CL) is one of the parameters that determine the molecular structure of starch molecules (Zhu, 2018). The size of the side branches will affect both crystallinity and properties like gelatinization, retrogradation, and pasting behavior in starch (Cai & Shi, 2010). FMS contains amylose having a typical chain length of 180–240 glucosyl units, around six chains in its molecule (Jideani *et al.*, 1996). On the other hand, amylose derived from pearl millet is said to have an average chain length of 260–270 glucosyl units, with each molecule having four strands, along with the typical level of polymerization ranging from approximately 1060 to 1250 (Gaffa *et al.*, 2004). Amylose contents in millet starches, analysis utilizing gel permeation chromatography categorizes amyloses into two groups: short-chain and long-chain 30. Comparing pearl and finger millet to foxtail and proso millet, which had ratios of 0.95 and 1.27 for long-chain amylose (LCAM) to Short-Chain Amylose (SCAM), respectively, and correspondingly greater SCAM levels, the former two exhibit a larger percentage of LCAM (Annor *et al.*, 2014a). The average pearl millet's chain count per molecule of amylose has been determined to lie in a narrow range between 3.9 and 4.8, close to the values of corn and rice starches (2.9–4.3) (Gaffa *et al.*, 2004). Finger millet amylose features longer glucan chains at the branch points when compared to the amylose from pearl millet, foxtail millet, and proso millet. For finger millet amylopectin, it is reported that the average chain length falls between 18 and 21 with internal chain lengths ranging from 5 to 6.3 and external chain lengths between 11.9 and 14 (Annor *et al.*, 2014b). Amylopectin fractions consist of unit chains that cluster to form the starch granules (Zhu, 2014b). The core is made up of little, branching units, so-called "building blocks" contributing to its amorphous nature, while the outwards part creates crystalline areas (Pérez & Bertoft, 2010). However, the clustering and organization of the fundamental components in millet amylopectin fractions have not yet been fully investigated; hence, the exact structure remains under investigation.

6.3 Physico-Chemical Properties of Finger Millet Starch

Finger millet is increasingly recognized for its numerous health benefits, including its high content of dietary fiber, essential minerals, and antioxidants, which contribute to improved overall health. The chemical composition and structural characteristics of FMS are important for its functional applications in various food products. Various techniques of starch isolation from finger millet include both the dry and wet milling procedures; both methods have further implications in the yield and quality of starch. Besides, physico-chemical properties of FMS include gelatinization, pasting behavior, and water absorption, which will be great regarding the suitability of FMS in various culinary uses. The key aspects concerning this are going to be elaborated herein, some proving the importance of FMS in food science and nutrition.

6.3.1 Surface Morphology

Important information on the changes that would have occurred at different stages of processing is obtained by Scanning Electron Microscopy (SEM). It allows observing the changes caused in millet starch, at the microstructure level, because of hydrothermal treatment, decorating, growing, and exploding (Jhan *et al.*, 2020). The morphological properties, microstructural aspects, and organization analyzed were obtained granules of FMS using SEM at an applied voltage of 15 kV (Gautam *et al.*, 2021). Dharmaraj *et al.* (2014), investigated the effects of processing on finger millet using SEM analysis. They found that the previously organized and structured arrangement of starch granules, along with other elements within the granule, shifted into a uniform and shapeless mass (Dharmaraj *et al.*, 2014). The SEM images of white finger millet-modified ultrasonicated starch (US) and annealed starch (AS), and native starch (NS) were constituted by different shapes such as polygonal, nonuniform, and oval shapes (Fig. 6.5). Minimum surface modifications are recorded in the starch treated with ultrasonication due to the poor gelatinization of millet starch. On the contrary, annealing (ANN) and ultrasonication treatments showed profound variations in white FMS morphology (Amarnath *et al.*, 2023).

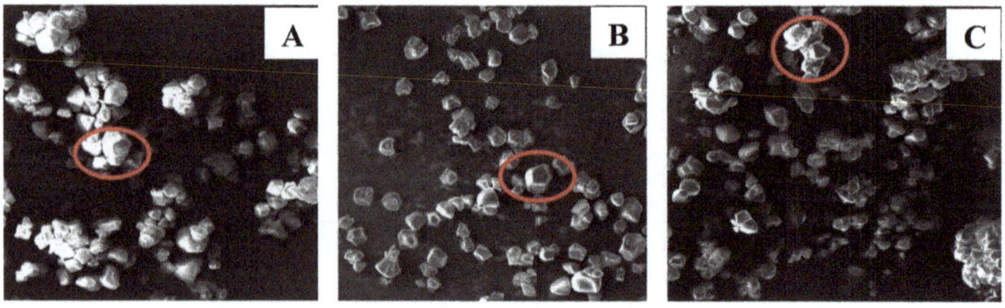

FIGURE 6.5
The SEM images of white finger millet-modified (A) native starch (NS), (B) annealed starch (AS), and (C) ultrasonicated starch (US) (Amarnath *et al.*, 2023).

6.3.2 Fourier transform infrared (FTIR) Analysis

FTIR analysis is a method used for identifying and characterizing chemical bonds in materials by assessing the infrared spectrum. This was performed at a resolution of 2 cm^{-1} between 370 and 4000 cm^{-1} (Amarnath *et al.*, 2023). Gautam *et al.* (2021) found that FMS indicated some distinctive peaks in its IR spectrum. The wide and clear peaks recorded between 3300 and 3100 cm^{-1} were ascribed to Oxide Hydroxide (OH) stretching vibrations owing to hydrogen bonding between hydroxyl groups. Further, a peak in the Carbon Hydroxide (CH) stretching of glucose units was associated with the 2900–3000 cm^{-1} range (Gautam *et al.*, 2021). The existence of moisture in the amorphous region of starch was shown by peaks seen between 1630 and 1660 cm^{-1}. Unique corresponding peaks were found at 1417 and 1001 cm^{-1}, corresponding to FMS due to C-H bending and C-O stretching vibrations. There was evidence of the glycosidic linkages in starch from the signals in the region of 860–770 cm^{-1}. Comparing the FTIR spectra of starch samples from different finger millet cultivars, some difference in the spectral pattern was noted. Peaks at a range of 450–900 cm^{-1} were assigned to the pyranoid ring-related skeletal vibrations. Amarnath and Collageus (2023) observed that the starch changes such as heating and ultrasound, both individually and in combination, did not alter the natural or modified starch's spectral patterns in white finger millet, showing that these treatments were purely physical processes (Fig. 6.6). The modified starches gave different peak intensities and transmittances. Relatively minute variations in the ratio of intensity of some

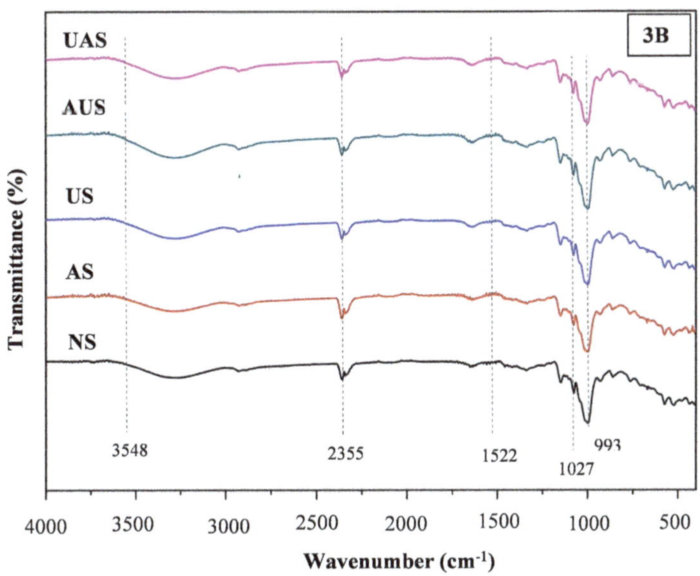

FIGURE 6.6
FTIR spectra of native and modified WFM starches (Amarnath *et al.*, 2023).

important wavenumbers in FTIR between modified and native white FMS point to a change in molecular structure and involve the ratio at 1022 cm^{-1} to 991 cm^{-1}, and 1047 cm^{-1} to 1022 cm^{-1} (Amarnath *et al.*, 2023).

6.3.3 Swelling and Solubility

Both the swelling index (SI) and water solubility of starch granules are indicators of their reaction to heat and water (Jia *et al.*, 2023). Table 6.2 depicts the changes in SI and water solubility resulting from various modifications. Among the various treatments, enzymatically modified FMS showed at 16.94%, the maximum solubility along with a pretty high swelling power (SP) of 3.18 g/g, reflecting its very high water absorption and swelling ability. In contrast, its counterpart high-pressure processing (HPP) treatment presented a minimal solubility of as low as 9%, although its SP was moderate, at about 3 g/g, showing poor SI degree. In a study, Amarnath *et al.* (2023) reported

TABLE 6.2

Swelling Power and Solubility of FMS

Type	Treatment	Solubility (%) at 90°C	SP (g/g)	References
FM	NS	14.85	3.53	Mahajan *et al.* (2021)
	AMS	10.70	3.45	
	HTMS	13.40	3.49	
	EMS	16.94	3.18	
WFM	NS	11	13.1	Sherin *et al.* (2024)
	HPP 200 MPa	5.1	10	
	400 MPa	7	10.9	
	600 MPa	3	8.8	
WFM	NS	–	14.91	Amarnath *et al.* (2023)
	CPI 170 V/10 min	–	12.32	
FM	170 V/20 min	–	12.11	Kaur and Annapure (2024b)
	170 V/30 min	–	12.16	
	200 V/10 min	–	12.02	
	200 V/20 min	–	12	
	200 V/30 min	–	12.53	
	230 V/10 min	–	11.93	
	230 V/20 min	–	13.98	
	230 V/30 min	–	14.02	

Abbreviations: FM: Finger millet, NS: Native finger millet starch, AMS: Acid-modified finger millet starch, HTMS: Hydrothermally modified finger millet starch, EMS: Enzymatically modified finger millet starch, HPP (600 Mpa): High-pressure processing, and CPI: Cold plasma–induced modification.

that different modifications did not alter the SI degree of native/modified starch at 50°C across all samples. Nonetheless, SP increased significantly, as the temperature increased from 50°C to 90°C, for changed starches. NS and FMS gave a SP of around 1.5 g/g at 50°C, which increased to 6 g/g at 90°C. At the intermediate temperature of 70°C, US starch exhibited a maximum SP of about 2.33 g/g, while another example gave values comparable to NS at 2.12 g/g. SP significantly increased at 90°C up to 6.54–6.58 g/g for modified starches, much higher compared with the NS of 2.31 g/g. It can be deduced that the SP of the modified starch increased either due to the uptake of water molecules in the starch's amorphous phase during ANN or as a result of strengthened hydrogen bonding among water molecules from ultrasonic treatment. Concerning solubility, both native and modified starches showed different behaviors at three temperatures: 50°C, 70°C, and 90°C. The modifications led to an increase in the solubility of white FMS as the temperature rose. The solubility of NSs was also shown to increase to 5% at 90°C. Temperature-related changes in intramolecular and intermolecular hydrogen bonding are responsible for these variations in solubility. Among the modified starches, NS had the lowest solubility and ultrasonic annealed starch the highest solubility over the whole temperature range (Amarnath *et al.*, 2023).

6.3.4 Gelatinization

Research has shown that the differences in starch gelatinization features occur not only across millet species but also among their genotypes (Zhu, 2014b). The Differential scanning calorimetry (DSC) analysis showed that to of different millet starches was in the sequence proso millet [T_0 = 68.4°C] > foxtail millet [T_0 = 66.7°C] > finger millet [T_0 = 63.9 C] > pearl millet [T_0 = 62.8°C]. Also, the starch enthalpy valu3es recorded in the millets were as follows: finger millet (13.2 J/g) > proso millet (13.1 J/g) > pearl millet (12.3 J/g) > foxtail millet (11.8 J/g) (Annor *et al.*, 2014a). These results are in sharp contrast to those obtained by Wu *et al.* (2014), who reported the gelatinization temperatures in a decreasing order of foxtail millet (69.7°C) > proso millet (68.7°C) > pearl millet (64.8°C) > Indian barnyard millet (64.7°C) > finger millet (63.4°C). The corresponding values of ΔH were found to be in the order of proso millet (5.2 J/g) > foxtail millet (4.0 J/g) > pearl millet (3.5 J/g) > Indian barnyard millet (3.3 J/g) > finger millet (2.9 J/g). These differences may be because of different varieties of millet as well as due to environmental and geographical conditions (Zhu, 2014b).

Amylopectin crystal quality and homogeneity are indicated by the starch melting temperature (T_c–T_0) range (Annor *et al.*, 2014a). A high proportion of amylopectin gives a broad range of melting temperatures. Amylopectin crystals that exhibit wide melting ranges tend to be less consistent, resulting in varying stability, and the opposite is also true (Ratnayake *et al.*, 2001). For instance, the amylopectin of finger millet presented the widest melting temperature range (T_c–T_0 = 10.2°C), spanning from 63.9°C to 74.1°C, which is broader than that of the amylopectin from other millets. Further, the amylopectin crystals found in pearl and finger millets display greater variability than those from foxtail and proso millets (Annor *et al.*, 2014a). Information for a similar kind of millet also seem to be quite divergent between the two studies; this may be because their regions of origin were different, as well as differences in shape, size, and starch granule composition, such as in amylose and amylopectin proportions (Zhu, 2014b).

The nature and extent of gelatinization occurring in the small millet starches are influenced by numerous factors including the ratio of amylose to amylopectin and the granular size of the starch. This can be expected even within varieties within a species. Starches that have high amylose contents tend to be more quick toward attaining a gelatinization temperature when compared with starches containing a lower amylose content or waxy starches (Koch & Jane, 2000). The gelatinization capabilities of starch are negatively correlated with the amylose content they contain (Cornejo-Ramírez *et al.*, 2018). Compared to tiny starch granules, larger ones need longer time to produce gels and reach greater gelatinization temperatures (Koch & Jane, 2000). Perhaps as a result of the tiny granules' increased potential for hydration and swelling (Brites *et al.*, 2008).

6.3.5 Retrogradation

When gelatinized starch cools, amylose and amylopectin molecules engage with water and with each other, leading to a more organized re-crystallization. Several parameters affect this retrogradation process, including water content, storage conditions and duration, amylose levels, the molecular structure of amylopectin, and the presence of non-starch elements such as lipids (Hoover, 2001). Retrogradation process can be accessed via different techniques, with DSC analysis and rheometry findings being used specifically for millet starch (Ojijo & Shimoni, 2007). A comparative DSC analysis exhibited that retrograded FMS indicated a lower onset temperature and an elevated ΔH when compared to pearl, foxtail, and proso millets (Annor *et al.*, 2014a). Nonetheless, this may not accurately represent the true differences among species, as a broader range of genotypes needs to be assessed. In another study, using rheological measurements, it was found that tapioca starch depressed retrogradation in FMS upon cooling, while adding xanthan gum to the mixture tended to increase retrogradation (Ojijo & Shimoni, 2007). Generally, much of the information about the effect of various factors on millet starch retrogradation and kinetics in respect to its morphological transformation remains to be known.

6.3.6 Textural Profile

The textural characteristics of starch gels, comprising consistency, cohesion, and firmness, were assessed by using the back extrusion method facilitated by a Texture Analyzer (Xing *et al.*, 2023). The hardness of the WFM starch gel was within the range of 0.13–0.31 kg/s. Among these values, the lowest hardness of 0.13 kg/s was recorded for US, while AS showed the highest hardness at 0.31 kg/s. Ultrasonication is a process that can disrupt the starch molecular structure, which negatively affects its gelation process, whereas ANN allows reorganization in the starch molecules, therefore resulting in higher hardness (Amarnath *et al.*, 2023). Ultrasonication treatment and retrogradation during ANN may have accounted for the increased hardness of Ultra-Sonication and Annealing (UAS). Addition of Guar Gum (GG) hydrolysate increased the gel hardness further, whereas native GG resulted in softer gels because it inhibited retrogradation of starch, hence yielding a more desirable texture (Mahajan *et al.*, 2021).

6.3.7 In Vitro Starch Digestibility

The digestibility of starch significantly impacts metabolic functions, influencing the speed of Digestive processes and thus impacting blood glucose levels (Xiao *et al.*, 2021). Starch can be divided into three categories: fast digestible starch, which is metabolized in around 20 min; gradually digested starch, which takes more than 20 min and up to 120 min to digest; and resistant starch (RS), which resists digestion for 120 min and thus enters the large intestine (Bangar *et al.*, 2021). The digestion and absorption of RDS in the small intestine are very fast; hence, blood glucose level increases rapidly. This rapid increase in blood sugar enhances the chance of developing CVD, obesity, and type 2 diabetes (Pranoto, 2022). Anti-nutritional agents, like tannins and phytic acid, may inhibit the proper action of digestive enzymes and decrease starch digestibility in finger millet. Traditional processing, involving soaking and fermentation, degrades the structure and reduces the content of such anti-nutritional factors, which enhances the digestibility of starch present in finger millet (Annor *et al.*, 2017). According to a study done by Onyango *et al.* (2020), native FMS has a digestibility rate of 72%, whereas cassava starch indicated a higher rate of 86%. The lower digestibility of FMS can be ascribed to different parameters, including the morphology of starch granules and their interaction with proteins, fiber, lipids, and polyphenols. Much of the RS in finger millet is not readily available to the enzymes for digestion. There is, therefore, a unique structure of the starch granules, with higher amylose content along with a densely packed crystalline formation, contributing to its resistance against enzymatic breakdown. Besides, phytochemicals associated with finger millet may impede the action of digestive enzymes due to its main constituent fiber, proteins, lipids, and polyphenols (Onyango *et al.*, 2020).

Removing lipids and proteins enhances the low glycemic index and low starch digestibility of FMS. In biscuits made from germinated finger millet and groundnut flour, there was a decrease in rapidly digestible starch, phytic acid, glycemic index, and starch hydrolysis index. Conversely, there was an increase in slowly digested starch, RS, total phenolic content, and antioxidant activity (Chinma *et al.*, 2022). In a specific in vitro study on the digestibility of finger millet, it was observed that while fermentation had a minimal impact on overall digestibility, it significantly increased the levels of slowly digestible starch (SDS) and reduced the amount of RS. This suggests that fermentation improves the digestibility of millet starch, likely by enabling microbial enzymes to break down amylose and alter the starch's crystalline structure. Indeed, fermentation makes the typically resistant crystalline structure of long-chain amylopectin more accessible to enzymes, resulting in a decrease in RS and an increase in SDS (Gupta & Gaur, 2024).

6.3.8 Pasting Characteristics

The pasting profile gives enough information about starch behavior in both the heating and cooling phases involved in its cooking. The starch properties were analyzed by using a Rapid Visco Analyzer (Bangar *et al.*, 2021). Numerous elements, including composition (such as amylose content), starch concentration, pH, cooking and cooling temperatures, and the presence of solutes like lipids and sugars, significantly impact the pasting behavior and characteristics of starches (Zhu *et al.*, 2020). Table 6.3 outlines the differences in pasting properties because of various modifications. In the case of WFM, AS treatment improves the interaction and reassembly of amylose and amylopectin chains, while US brings about significant alterations in pasting properties by increasing amylose levels and causing disruptions in the molecular chains (Amarnath *et al.*, 2023).

The peak temperatures range between 94.85°C and 94.92°C and are very similar for both native and modified starch. The peak viscosity of modified starch is significantly higher compared to that of NS. Further, the pasting temperature for MS was 74.53°C, while that of hydrothermally modified starch is 74.46°C, both of them are far different from the NS which is 73.41°C. Whereby viscosity of holding in NS showed a big difference of 0.39 cP, while in MS and hydrothermally modified starches, the differences were relatively small, being 1.38 cP and 1.50 cP, respectively (Balakumaran *et al.*, 2023). These alterations were marked by a small increase in the pasting temperature of native FMS gels, alongside decreases in both peak viscosity and breakdown viscosity. Further, there was a significant rise in breakdown viscosity, a drop in final viscosity, and a pronounced increase in setback viscosity (Kumar *et al.*, 2024).

6.3.9 Thermal Analysis

Thermogravimetric analysis (TGA) is a technique usually utilized in the assessment of the thermal stability of materials. TGA provides useful information on the maximum temperature at which material processing may occur (Gautam *et al.*, 2022). The thermal stability of FMS in a nitrogen

TABLE 6.3

Pasting Properties of Native and Different Modified Finger Millet Starches

Sample	PV (Pa.s/cP)	TV/HV (Pa.s/cP)	BDV (Pa.s/cP)	FV (Pa.s/cP)	SBV (Pa.s/cP)	P time (min)	P temp (ºC)	PT (ºC)	References
NS	2.038	3.236	1.594	3.746	−1.661	–	98.85	73.41	Balakumaran *et al.* (2023)
HS	2.962	2.825	1.474	4.254	−1.343	–	98.89	74.46	Balakumaran *et al.* (2023)
MS	2.977	2.658	1.599	3.993	−1.040	–	94.72	74.53	Balakumaran *et al.* (2023)
NS	3.354	–	1.658	5.232	1.878	6.18	–	–	Kaur and Annapure (2024a)
CPI 170 V/10 min	3.059	–	1.276	4.844	1.785	6.43	–	–	Kaur and Annapure (2024a)
CPI 170 V/20 min	3.094	–	1.272	4.778	1.684	6.59	–	–	Kaur and Annapure (2024a)
CPI 170 V/30 min	2.993	–	1.105	4.756	1.763	6.72	–	–	Kaur and Annapure (2024a)
CPI 200 V/10 min	3.052	–	1.096	4.844	1.792	6.69	–	–	Kaur and Annapure (2024a)
CPI 200 V/20 min	2.899	–	0.932	4.402	1.503	6.92	–	–	Kaur and Annapure (2024a)
CPI 200 V/30 min	2.518	–	0.944	3.123	0.605	6.99	–	–	Kaur and Annapure (2024a)
CPI 230 V/10 min	2.943	–	0.866	4.309	1.366	6.99	–	–	Kaur and Annapure (2024a)
CPI 230 V/20 min	1.897	–	0.367	1.429	1.468	6.66	–	–	Kaur and Annapure (2024a)
CPI 230 V/30 min	0.055	–	0.032	0.048	0.007	6.38	–	–	Kaur and Annapure (2024a)
NS	9237	–	5081	7421.5	1801.5	–	–	74.14	Sherin *et al.* (2024)
HPP 200 MPa	11,355	–	5898	8676	2698	–	–	74.15	Sherin *et al.* (2024)
HPP 400 MPa	10,520	–	5172.5	8804	2083	–	–	73.73	Sherin *et al.* (2024)
HPP 600 MPa	12200	–	4366	12,690	−1771	–	–	74.49	Sherin *et al.* (2024)

Abbreviations: NS: Native finger millet starch, HS: Hydrothermally modified starch, MS: Microwave treated starch, CPI: Cold plasma–induced modification, and HPP: High-pressure processing.

atmosphere was investigated by testing around 10 mg of the specimens within the temperature range of 30°C to 600°C. Thermal analysis showed the mass loss of FMS as a function of increased temperature in an inert environment. According to Yogananda *et al.* (2021), the thermal degradation temperature (Td) for FMS was within a range of 282.77°C to 287.85°C, whereas according to Gautam *et al.* (2021) decomposition of starch occurred in two distinct steps. The temperature sequence began at 35°C, then increased from 75°C to 120°C, during which the extracted FMS was evaporating its unbound moisture; among them, it was noticed that some starch samples showed low values of unbound moisture; further, with the temperature rising from 120°C to 270°C, slight starch structures and bound moisture became noticeable. As the temperature advanced from 270°C to 350°C, heat-induced degradation of starch occurred, various byproducts-gas carbon dioxide, acetaldehyde, water, gas carbon monoxide, furan, and 2-methyl furan (Gautam *et al.*, 2021).

6.4 Approaches to Modify Finger Millet Starch

A summary of different approaches for the modification of FMS is presented in Table 6.4.

6.4.1 Chemical Modification

6.4.1.1 Acid Treatment and Oxidation

Starch undergoes chemical modification through the application of agents such as sulfuric acid (H_2SO_4), nitric acid (HNO_3), HCl, and phosphoric acid (H_3PO_4), all at temperatures that are lower than their gelatinization thresholds (Mahajan *et al.*, 2021). This treatment cleaves the glycosidic bonds of amylose and amylopectin. After treatment of starch, it alters its functional and structural properties (Sinhmar *et al.*, 2023). Acid hydrolysis is a chemical process initiated by the presence of a positively charged species known as the hydronium ion (H3O+). This ion plays a crucial role in breaking down glycosidic bonds, specifically the α 1 $\rightarrow$ 4 linkage found in polysaccharides. The mechanism begins with the hydronium ion focusing its attention on the oxygen atom within the glycosidic bond. This interaction facilitates the protonation of the oxygen, making the bond more susceptible to nucleophilic attack by water molecules. As a result, this process leads to the cleavage of the glycosidic bond, ultimately yielding the component monosaccharides.(Bangar *et al.*, 2021). After undergoing acid modification, FMS exhibits an intrinsic viscosity of 131 mL/g at 24°C, an iodine binding capacity of 4.93%, and an amylose equivalent of 25.3%. Acid modification results in a decrease in the intrinsic viscosity of FMS because of partial hydrolysis and

TABLE 6.4

Summary of the Different Approaches to Modify Finger Millet Starch

Type	Process	Key Effects	Benefits	Limitations	Considerations
Chemical	Acid Treatment and Oxidation	Disruption of glycosidic bonds; reduced viscosity	Enhanced shear stability; alters functional properties	Viscosity reduction varies by acid type; potential degradation	Requires controlled conditions for acid concentration
	Hydroxypropylation	Decrease in turbidity; increased peak viscosity	Improved pasting properties; reduced retrogradation	Varied effects based on starch type	Analyze starch genotype for optimal results
Physical	Hydrothermal Treatments	Alters crystalline arrangement; restructures polymer chains	Environmentally friendly; improves digestibility	Requires precise control of temperature and moisture levels	Different techniques have varying effects
	Microwave Treatment	Improved gel hydration; increased amylose content	Enhanced nutritional application; low energy requirement	Optimization required for commercial application	Monitor power and duration closely
	Autoclave Treatment	Swelling of starch granules; altered crystalline form	Enhanced water-binding; improved dough formulation	May lead to structural deformation	Monitor pressure and temperature closely
	UHP	Gelatinization and increased SP	Reduced energy usage in baking; modified thermal characteristics	Sparser research on millet starches using UHP	Parameters like temperature and duration are crucial
	Cold Plasma	Changes in solubility; reduced viscosity	Environmentally friendly; alters crystallinity and functionality	Extreme conditions can severely affect structural integrity	Variable effects based on voltage applied
	Ultrasonication	Improved water solubility; microstructural changes	Denser, more uniform structure	Potentially reduces gel formation	Adjust treatment intensity and frequency for best results
	DHT	Increased solubility; improved film properties	Effective for producing edible films and coatings	May reduce amylose content	Analyze impact on mechanical properties
Enzymatic	Enzymatic Modification	Reduced viscosity; improved light transmittance	No harmful chemicals used; stable structure; reduced syneresis	Limited to specific enzyme applications	Monitor enzyme concentration and type for efficiency

Abbreviations: UHP: Ultra-high pressure and DHT: Dry heat treatment.

depolymerization; stronger acids such as HCl and HNO_3 result in more deterioration in comparison to less aggressive acids such as H_3PO_4 and H_2SO_4. Similarly, the molecular weight of FMS reduced by 6% with H_3PO_4, 25% with H_2SO_4, 42% with HCl, and 39% with HNO_3) (Singh & Ali, 2008). In a similar vein, starch oxidation represents a different method of chemical modification that utilizes oxidizing agents such as sodium hypochlorite (Dimri *et al.*, 2023). This process involves the substitution of hydroxyl groups with carboxyl (COOH) and carbonyl (C=O) groups, leading to changes in the molecular structure and various characteristics of the starch (Anwar *et al.*, 2020). The viscosity of the paste decreased as a result of FMS oxidation, which was linked to the disruption of its granular structure and reduced intermolecular forces, thereby enhancing shear stability. Despite the modification, transparency remained unchanged, likely because the rearrangement of the oxidized starch molecules did not promote improved light transmission (Mahajan *et al.*, 2021).

6.4.1.2 Hydroxypropylation

Starch's hydroxyl groups are capable of reacting and can be replaced by different functional groups, including hydroxypropyl groups. FMS hydroxypropylation principally impacts the sixth carbon position, based on Carbon-13 nuclear magnetic resonance (NMR) spectroscopic results. This modification leads to a decrease in setback value, pasting temperature, and turbidity, as well as reduced syneresis and retrogradation, while simultaneously increasing the peak viscosity during the pasting process (Lawal, 2009). Earlier research indicated that the effect of hydroxypropylation on peak viscosity during pasting varies depending on the type of starch used (Gunaratne & Corke, 2007). Therefore, altering millet starch from various genotypes or species is likely to yield diverse properties. In addition to hydroxypropylation and acid treatment, it could be valuable to explore other chemical modifications like succinylation, acetylation, octenyl succinylation, carboxymethylation, as well as methods like oxidation, cross-linking, and alkaline treatment.

6.4.2 Physical Modification

6.4.2.1 Hydrothermal Treatments

Heat moisture treatment (HMT), sometimes referred to as hydrothermal treatment, is a process of physical modification that makes use of regulated heat and moisture. This technique typically operates under moisture levels of less than 35% (w/w) and requires exposure to temperatures that exceed gelatinization thresholds, lasting anywhere from 15 min to 16 h (Arns *et al.*, 2014). This process partially alters the crystalline arrangement, influencing how the polymer chains interact with each other. As a result, the double-helical configuration breaks down, leading to the eventual reorganization of the disrupted crystals (Gunaratne & Hoover, 2002). The impact of HMT on starch enzyme digestibility is primarily determined by several factors: (i) the origin of the starch, (ii) the level of moisture present, (iii) the thermal conditions, (iv) treatment time interval, and (v) the intramolecular and intermolecular interaction between different components. Indeed, this encompasses interactions among amylopectin molecules, interactions among amylose molecules, and interactions between amylose and amylopectin (Pratiwi *et al.*, 2018). Different hydrothermal processes, like HMTs, ANN, and quenching (QUE), have been documented concerning FMS (Adebowale *et al.*, 2005). HMT typically takes place at temperatures ranging from 100°C to 130°C, with moisture levels kept below 35%. On the other hand, ANN involves heating starch to a water content of approximately 40–65% while maintaining temperatures below the point of gelatinization (Hoover, 2010). QUE is the process of swiftly cooling an overheated starch slurry to significantly lower temperatures, such as –180°C (Premavalli *et al.*, 2005). Hydrothermal treatments promote the interaction and reassembly of amylose and amylopectin chains within the granules. At the same time, ANN and HMT improve the structural deficiencies of the crystalline areas in these granules. Additionally, both HMT and ANN increase the gelatinization temperatures and ΔH in FMS (Premavalli *et al.*, 2005), while QUE reduced the gelatinization temperatures (Premavalli *et al.*, 2005). Although the molecular processes of HMT and ANN are well comprehended, the mechanisms of QUE still require further investigation. Physical modifications seem to be a more environmentally friendly approach to altering starch compared to chemical and enzymatic methods, making them healthier for consumers.

6.4.2.2 Microwave Treatment

Generally, the dielectric heating and electromagnetic polarization phenomenon act from the very beginning of the application on the starch. It is the modification that will alter physical and chemical changes in starch, probably improving the functionality of NS (Zheng *et al.*, 2020). Microwaves constitute a kind of non-ionizing electromagnetic radiation characterized by a certain frequency spectrum used by microwaves ranging from 300 GHz to 300 MHz. These waves have varying natures of penetration depending on the material it is interacting with, raising the temperature of such materials by interacting with the molecules using microwave energy (Yi *et al.*, 2024). NS lacks some desirable properties of starch such as shear stability and low-temperature gelatinization. Modification into desired functionality and structure in food products from starch is done through physical, chemical, enzymatic, and hybrid methods. Observations have indicated that hydrothermal and microwave modification techniques drastically affect the functionality of starch, extending the nutritional application of FMS. Such modification affects the natural starch's thermal stability, physiochemical properties, functionality, and microstructural characteristics. However, the gel hydration capacity of starch of the finger millet improves, but oil absorption capacity remains unchanged. These improved hydration qualities of gel can become useful in the preparation of films and other applications The application of microwave treatment enhanced the levels of amylose and resistant starch in the FMS, likely due to the gelatinization process that disrupted the starch's crystallinity. As a result, this alteration made it easier for amylopectin to convert into amylose; hence the leaching (Zhi *et al.*, 2022). Microwave treatment hardly affected the chemical composition of native white FMS and water activity; however, the pH may have been reduced. On the other hand, microwave treatment enhanced the gel hydration characteristics by increasing the Least gelation concentration (LGC) of white finger millet from 13.63% to 16.41%. Hydration and swelling of starch granules are the major factor in gel formation, which mainly happens in the amorphous regions of the starch (Balakumaran *et al.*, 2023). First and foremost, control of process variables including time and power has to happen prior to commercial application of microwave treatment. The second most critical moot issue will be the determination of the way in which this technique affects the properties of millet starch. Later on, this technique may be employed in an inexpensive way for producing high resistant starch for various functional food applications.

6.4.2.3 Autoclave Treatment

Food items undergo elevated pressure and temperature when treated in an autoclave. This autoclaving process generates a pressure environment that promotes the absorption of water in the amorphous regions of starch granules. As the treatment pressure escalates, the crystallization area of the amorphous layer within the starch granules enlarges, causing the amylopectin double helices to realign (Bravo *et al.*, 1998). In this process, the nature of starch can be affected in more than one way, especially for millet starch. Heating and pressure could result in swelling of the starch granules and loss of crystalline form. The improvements in the solubility and reduction in viscosity are achieved with the distortion of the crystalline structure. In gelatinization, the temperature increases gradually, while the pressure is kept at a constant level; therefore, the treatment of autoclaving can distort the crystalline structure to result in the formation of deformed crystals (Yang *et al.*, 2016). During the retrogradation process, the processed starch might, upon cooling, acquire a more crystalline and ordered structure that would enhance hardness and reduce its water solubility. In the cross-linking process, when heated in an autoclave, starch molecules may experience the possible formation of covalent bonds that make them more stable and digestible. Functional modification during treatment in an autoclave can alter some important functional properties of millet starch, for instance, its ability to bind water, dough formulation, and pasting properties.

6.4.2.4 Ultra-High Pressure (UHP)

Research on the application of high-pressure processing (HPP) or UHP for modifying millet starches remains sparse. UHP is an environmentally friendly, non-thermal food processing technology. It has been utilized in studies involving dairy items, purees, shellfish, fruit juices, and meat. It has been employed in altering non-covalent chemical bonds, a feature essential for providing starch with basic properties. The degree of swelling and starch leaching can increase with pressure, upon the temperature conditions prevailing. Higher values of pressure have, at times, achieved distortion in the starch granule, which leads to its gelatinization and at times to the lowering in birefringence of these granules. Pressure can significantly modify thermal characteristics, pasting behaviors, and the amount of resistant starch. More so, this approach could potentially prevent starch retrogradation and transform type A starch polymorphism into type B polymorphism. The gelatinization of starch induced by UHP in customized and functional foods presents numerous potential uses. The degree of this gelatinization is influenced by various parameters, including temperature, pressure, type of starch, concentration, exposure period, and the nature of the medium used (Castro *et al.*, 2020).

Utilizing starches that have been treated with HHP in baked products provides the benefit of a reduced gelatinization onset temperature, which could result in lower energy usage in the baking sector (Grgić *et al.*, 2019). Results from these research projects exhibited that less heat is required to disrupt the intermolecular bonds in starch subjected to elevated pressure. In a study conducted by Sherin *et al.* (2024) on WFM starch, it was reported that this treatment affected various color metrics of the modified starch, including L*, a*, b*, and ΔC, while also causing a significant decline in paste clarity as time progressed. The concentration for gelation was found to be between 8% and 14%. A temperature as high as 90°C resulted in increased water solubility and swelling index of the starch. Starch processed at 600 MPa exhibited an ordered microscopic pattern with a B-type crystalline structure and distortion on the surface, which was indicative of gelatinization. This pressure also shifted FTIR bands at 3330, 2358, and 997 cm^{-1}, reflecting the loss of crystallinity. FMS exhibited a rise in the amount of amylose by about 5% without showing any major modification in its chemical composition. HHP increases water and oil solubility, absorption, and specific viscosity and further increases syneresis. Low-pressure effect improved freeze-thaw stability. In starch modified by HPP, the result showed a reduction in setback and breakdown viscosities, and signs of stronger gels with increased resistance to retrogradation. As such, the pressure changed amylopectin; thus, short-range ordered crystallinity decreased. Although the granular structure remained intact, the application of intense pressure resulted in agglomeration, modifying the amylose-amylopectin configuration (Sherin *et al.*, 2024).

6.4.2.5 Cold Plasma

Cold plasma (CP) modification of millet starch changes its physico-chemical and functional attributes via a low-temperature plasma methodology. Millet starch will be exposed to CP during this process, wherein free radicals and ions can be developed through the dissociation of gaseous atoms and/or molecules by applying energy. Such reactive species attack starch molecules chemically in the form of breaking bonds, thereby changing solubility, crystallinity, molecular weight, and functionality (Sun *et al.*, 2021; Zhu, 2014a). CP technique is gaining traction among researchers. When starch is exposed to CP treatment, it can be altered via different processes like cross-linking, depolymerization, plasma etching, and the introduction of functional groups, resulting in notable transformations in its main features (Singh *et al.*, 2022). The application of CP treatment to FMS led to a wide range of critical changes, including reduction of pH, intrinsic viscosity, zeta potential, molecular weight, amylose content, and gelatinization temperature, all of which suggest depolymerization has occurred (Kaur & Annapure, 2024a). Confirmation of increased carboxyl and carbonyl groups was done through FTIR analysis, where both affected syneresis and gel hydration. Also, light cross-linking for 200 V for 20 min and 230 V for 10 min showing great interest in C2 and C6 position sites was observed. On the contrary, the extreme conditions lead to the extreme reduction of viscosity up to 99.81%—those being 230 V for 20 to 30 min. Although there was a decrease in structural integrity, the strength of the gel enhanced at voltages of 170 V and 200 V owing to the inclusion of functional groups. CP treatment caused changes both physically (through etching) and chemically (by generating functional active groups and promoting different process like oxidation), with these effects being affected by variables such as polymer chain length and its molecular weight. This process increases the strength and stability of gels, positioning CP treatment as a viable and environmentally friendly technique for altering starch characteristics for use in large-scale applications (Kaur & Annapure, 2024b).

6.4.2.6 Ultrasonication

Ultrasound, an innovative physical technology, is utilized for the alteration of starch. The high-intensity ultrasound can produce mechanical stress and acoustic cavitation inside the starch particles when it is scattered with more than 1 W/cm^2 intensity, enabling an improvement in water solubility and structural modification of NS granules. Such modifications would improve crystallinity, gelatinization, retrogradation, viscosity, and swelling behaviors (Zhu, 2015). On the other hand, tri-frequency ultrasound causes greater damage to starch because of its high rate of cavitation and sudden implosion of bubble owing to the substantial cavitation yield and quick bubble implosion (Li *et al.*, 2019). In their 2019 study, Hu and associates investigated how single and dual-frequency ultrasound influences potato starch, which has larger granule sizes, alongside millet starch, known for its smaller granule sizes. They confirmed that potato starch exhibited greater sensitivity compared to millet starch, as the latter displayed only slight alterations in response to ultrasound application (Hu *et al.*, 2019). The highest viscosity experienced a reduction of up to 3.9% for millet starch, while the relative crystallinity diminished by 1.68%. It was noted that the SP increase at 25°C diminished when assessed at 60°C for both types of starch granules. According to the author, ultrasound treatment significantly affected the host section of starch with larger granules over the smaller granule starch and increased temperature for developing the process for modification. In 2021, Yadav and his team studied the impact of using ultrasound to hydrate finger millet. Their observed findings exhibited that ultrasound treatment improved the structural characteristics, the hydration duration, and enhanced both water solubility and absorption capability (Yadav *et al.*, 2021). US from WFM improved water absorption and solubility because of the breakdown of starch granules resulting from shear stress generated by cavitation, which enhances holding and the penetration of more water inside the pores of the granules. Additionally, it lowered the relative crystallinity of the starch structures and decreased the firmness of the starch gel. On the other hand, the cleaving of the starch molecules during the ultrasonication process negatively impacts gel formation.

Microscopic analysis indicated notable alterations in the microstructure and morphology of the modified starch, including the presence of voids and fissures (Amarnath *et al.*, 2023). The hydration through sonication increased solubility and the water absorption ability of FMS. X-Ray Diffraction (XRD) results for starch extracts revealed an increase in particle crystallinity. Field Emission Scanning Electron Microscopy (FESEM) images demonstrated that hydration through sonication leads to a reduction in particle size and more uniform dispersion of FMS particles (Yadav *et al.*, 2021). Diverse preparation techniques have varied effects on the morphology and shape of starch granules. However, the resistant starch generated through ultrasonic treatment demonstrates a consistent B-type crystallinity arrangement. The sedimentation value of finger millet-resistant starch prepared with ultrasonic assistance Ultrasonic Assistance (UAP-FRS) gradually rises over time, achieving a peak of 16.7%. This increase is attributed to the breakdown of starch granules and the development of a denser, more uniform structure that improves sedimentation. This starch requires the greatest amount of energy for complete gelatinization ΔH, 145 J/g, since the ultrasonic treatment enhances strong intermolecular and intramolecular bonding in this starch, thereby increasing its thermal resistance and stability (Wang *et al.*, 2022). Scaling up this method is consequently suggested by the authors as an alternative to conventional hydration methods of millet processing.

6.4.2.7 *Dry Heat Treatment*

The thermal process known as dry heat treatment (DHT) is applied to starch in a dry environment, altering its physico-chemical properties while leaving the granule structure intact (Pandey *et al.*, 2023). DHT of FMS increases its solubility and decreases amylose content, ash content, swelling, and moisture content. Such modifications are related to DHT-induced changes in the starch molecular microstructure, morphology, and thermal pattern because of a reduction in crystalline regions. Thin films produced from DHT-treated FMS exhibit greater uniformity and continuity, lacking the cracks and voids found in films derived from unaltered FMS. Such films also showed improved breathability, higher solubility in water, and excellent physico-mechanical attributes. In conclusion, DHT proves to be a highly efficient method for enhancing the physical and chemical properties of high-amylose FMS, making it suitable for the creation of edible packaging films or thin coating layers (Gautam *et al.*, 2023).

6.4.3 Enzymatic Modification

Because enzymatic starch modification uses no hazardous ingredients, breaks down starch granules into tiny holes, and decreases viscosity and water retention, it is preferred over chemical techniques (Zheng *et al.*, 2022). The process induces the production of extra amylose chains, which restrict the uptake of water and hence decrease the ability of FMS to swell. The modified FMS showed enhanced light transparency and decreased syneresis as a result of enzymatic changes (Punia *et al.*, 2021). The freeze-thaw stability, color value pasting behavior, paste clarity, and other properties of FMS processed enzymatically were comparable to those of starch treated with acid. The greater stability of starch structure leads to better characteristics, which may be attributed to the strengthening of both intermolecular and intramolecular connections within starch chains. This feature is very useful when making pastes and jellies, when it is necessary to maintain a uniform consistency (Mahajan *et al.*, 2021). The alteration process lengthens the amylose chains in FMS, which decreases their ability to absorb water and expand. For modified FMS, enzymatic modification achieves maximal transmittance while reducing syneresis (Bora *et al.*, 2019).

6.4.4 Other Modification Processes

6.4.4.1 *Cross-Linking*

Crosslinking may be done with various food-safe cross-linking agents: sodium tripolyphosphate, sodium trimetaphosphate, epichlorohydrin, and phosphorus oxychloride. This kind of starch modification, because of it, causes a decrease in the swelling of granules and the formation of a new link in which resistance to the structure and thermal decomposition is imparted.

Many of these have the added effect of improving overall stability. The degree of these effects is contingent upon many different factors (Bhatt *et al.*, 2023). Cross-linked starch is considered useful ingredient in health-conscious foods because it resists normal digestion degradation and its susceptibility for degradation is diminished (Bojarczuk *et al.*, 2022). Sharma and his team explored the combined impact of cross-linking agents on the features of films made from FMS. The research revealed that employing gelatin and succinic acid for cross-linking impacts the moisture content in the films, with those linked by succinic acid showcasing superior tensile strength (Sharma *et al.*, 2023). In FMS-based packaging films, elevated concentrations of cross-linking agents enhance both Young's modulus and tensile strength, but lead to a reduction in elongation at break. Although intramolecular and intermolecular cross-linking boosts both Young's modulus and tensile strength, it restricts the movement of gelatin molecules, consequently diminishing the films' flexibility and their ability to elongate before breaking.

6.4.4.2 Esterification and Etherification

Esterification modification entails the integration of new active groups, like succinic anhydride, into starch, leading to modifications to its structure and subsequent variations in its characteristics (Gupta & Gaur, 2024). Esterification breaks down the orderly structure of starch and hydrogen bonds, while facilitating the penetration of water to produce a paste with an increased viscosity. Selection of the type of esterification must be made with due care to develop the right functionality in targeted food applications (Sinhmar *et al.*, 2023). In contrast to succinylated starch, the majority of altered millet starches demonstrate lower pasting temperatures, suggesting that they gelatinize more readily. The pasting temperature of succinylated starch, however, is elevated because of microstructural and organization breakdown occurring during the modification process, which raises the energy needed for gelatinization (Muñoz-Llandes *et al.*, 2023). FMS was etherified using propylene oxide, resulting in a hydroxypropylated derivative that contained 52.4% starch based on dry weight. That a higher volume of propylene oxide resulted in a higher MS. Hydroxypropylation also improved FSC within a temperature range of 30°C–90°C due to the penetration of water into the amorphous regions, which reduced turbidity by prevention of re-formation of intermolecular and intramolecular bonds, increased peak viscosity by reducing the pasting temperature of the starch molecules and lowered the setback value. It showed that the hydroxypropylation occurred mainly in the sixth carbon of the anhydroglucose unit by means of carbon-13 NMR (Lawal, 2009). Incorporating succinic acid into FMS films has been noted to enhance properties such as breathability and thermal stability while also improving the overall features and visual appeal of the films. More so, the inclusion of this element resulted in enhanced strength, density, permeability, and surface texture of the films in comparison to other film varieties (Sharma *et al.*, 2023).

Starches generally altered through etherification via hydroxypropylation commonly utilize epoxide compounds like ethylene oxide and propylene oxide (Paramitasari *et al.*, 2024). Modifying FMS through oxidation enhances its functional attributes by reducing paste formation and retrogradation, boosting shear stability, and improving resistance to freezing and thawing, all while maintaining desirable textural properties without demonstrating greater clarity (Bangar *et al.*, 2024). Hydroxypropyl changes the granular architecture of FMS, weakening the intermolecular bonds and increasing its solubility. Such changes result in better flow characteristics and heightened thickness or viscosity (Thakur *et al.*, 2023).

6.4.4.3 Acetylation

This widely adopted modification technique involves blending a starch mixture with acetic anhydride (Mehboob *et al.*, 2020). The traits of starch that have been modified through the incorporation of hydroxyl and acetyl groups into its chains are influenced by the degree of substitution (DS). The DS degree can be influenced by various parameters, including pH levels, the reaction environment, the type and quantity of reactants, treatment time, catalysts, temperature, and additional conditions such as high-pressure or microwave processing (Ashogbon, 2021). Starch undergoes more vigorous acetylation in its dry form, as greater access to acetylating agents and the lack of water contribute to this enhanced reaction, giving a DS of about 2.9. In wet conditions, for example during cooking or processing where water is present, it limits the degree of acetylation and hence gives a lower DS of about 0.2. The fact that the DS is different in the case of dry and wet conditions not only points out the strong impact of surrounding circumstances on acetylation but also on the features of modified starch. Food Additives Expert Committee of WHO and FDA recommendations establish that the content of acetyl groups in modified starch obtained either with vinyl acetate or acetic anhydride should not exceed 2.5% (Bangar *et al.*, 2021).

6.5 Functional Properties of Modified FMS

Traditionally consumed as whole grain, modified FMS has emerged as a versatile ingredient in various food applications. This modified starch is subjected to physical or chemical treatments that enhance its functionality, making it suitable for a wide range of culinary uses, from thickening agents to fat replacers. Apart from functional attributes, modified FMS also contributes significantly to dietary improvements and overall health. The gelatinization and pasting properties of modified FMSs were superior to those of its NS (Shahidi & Chandrasekara, 2013). From the practical point of view, gelation and paste properties are imperative in sauces, soups, and gravies, as it is the starch that usually confers the proper textural and viscous feel. One of the primary advantages of modified FMS over its natural starch is its resistance against extreme variations in pH and temperature. This attribute is ideal for products subjected to different conditions of processing, such as canned foods and bakery items. Modified starch can retain functional integrity to ensure consistent product quality (Saleh *et al.*, 2013; Shahidi & Chandrasekara, 2013). Modified FMS can function as a fat replacer in low-fat or reduced-calorie food formulations. It can contribute to a reduction in overall calorie content by providing creaminess and the sensation of fat, while still maintaining desirable organoleptic properties and thus aligning with trends toward healthier eating (Park *et al.*, 2018). More so, the addition of modified starch can improve the texture of gluten-free products and make them more acceptable for consumption. Modified FMS, being one imitation for gluten properties, improves bread structure, pastry, etc., to cope with demand due to the increase in gluten-free requirements. Regarding water-holding capacity, modified starch shows excellent characteristics that may help in improving moisture retention of various food products. This is particularly desirable in baked goods, as this can lengthen the life of the food and preserve freshness (Dhanushkodi *et al.*, 2023).

Finger millet is composed of essential amino acids, vitamins, minerals, and dietary fiber. These qualities are retained in the modified starch, hence providing healthful additions to other starches that are more refined. Therefore, the modified FMS has great value in being added to diets for balance. There is an associated improvement in blood glucose control with the consumption of finger millet and its modified starch. This is partly because of the dietary fiber present, which helps to slow the absorption of glucose, hence blood sugar levels are controlled. Modified starch of finger millet maintains phytochemicals like antioxidants that reduce oxidative stress and inflammation in the body and hence could have a beneficial effect on cardiovascular health (Vedamanickam *et al.*, 2020). The high fiber content in finger millet contributes to good digestive health by ensuring regular bowel movements and avoiding constipation. Modified starch retaining its fiber content will contribute positively toward gut health for a healthy microbiota (Lu *et al.*, 2009). Moreover, foods containing added modified FMS can reduce hunger and lower overall food intake without affecting the palatability or textural characteristics of the food, and this should be useful for effective weight control (Dhanushkodi *et al.*, 2023). In conclusion, modified FMS offers a unique combination of functional properties and health benefits, making it an attractive ingredient in the food industry. As awareness of the benefits of whole grains continues to grow, modified FMS stands out as a promising alternative that aligns with contemporary dietary preferences for healthier, sustainable food options.

6.6 Applications of FMS

Starch is utilized in different food-related applications globally. Millet starch can be found in its native state or a processed or modified form for diverse culinary uses. The techniques employed to modify starch or change its functionality and various special characteristics, allow for broader utilization in food products. The techniques utilized for starch treatment play a particular role in altering the structure of starch, influencing its resistance to digestive enzymes. Modified starches serve various purposes, including meat applications, acting in baked products such as thickeners and binders, and snack seasonings. They also function as fat replacers in flavor encapsulators, ice cream, gelling agents in gels and gums, as well as emulsion stabilizers in drinks and juices. More so, as a means of delivering drugs in tablet formulations, starches that have been acid-modified, pregelatinized, acetylated, and esterified are commonly utilized. The application of FMS in various sectors is shown in Fig. 6.7.

6.6.1 Resistant Starch as a Functional Ingredient

It is the dry heat technique that contributes to resistant starch formation. Dry heating under acidic environment causes millet starch to form dextrin-a process called pyrodextrinization. Under these conditions, the digestibility

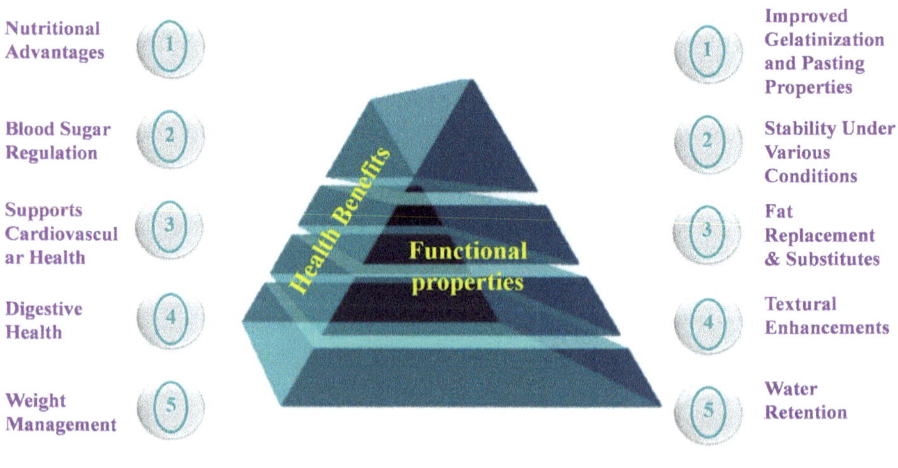

FIGURE 6.7
Functional properties and health benefits of modified finger millet starch.

of the starch is reduced due to an increase of dextrin levels and hence forms resistant starch. Sun and his team showed that in the case of double-modified starches, the preparation by acid hydrolysis in combination with HMT resulted in higher RS. RS are important, for they affect health through not being digested in the small intestine (Sun *et al.*, 2015). They undergo metabolism in the large intestine into smaller fatty acids, which have lower molecular weights and thus can be used in the treatment of cancer, obesity, diabetes, and several CVDs. Resistant starch can increase with a combination of heat treatment, chemical treatment, and ultrasound in millets. This helps in the production of gluten-free fiber-rich products (Babu & Mohan, 2019). Of the finger millet-resistant starches produced through ultrasound, ultrasonic-autoclaving, and enzymatic techniques, the highest levels of solubility and swelling capacity were noted. Thus, this starch is suitable for use in fried food applications. On the contrary, ozone-treated modified starch indicated high resistance to digestion with a low sedimentation value and is recommended for the preparation of functional beverages (Wang *et al.*, 2022). Kaimal *et al.* (2021) reported that research into the impact of HPP on RS in millet, as well as the influence of dietary elements, is scarce. Thermal and chemical methods of starch modification enhance starch resistance more significantly compared to alternative techniques There is still remarkable research needed on non-thermal techniques for modifying starch to boost RS levels. The RS from millet finds its many applications in bakery products, convenience foods to traditional ready-to-eat preparations. Millet starch may contain RS in very low to very high concentration. Millets that have low resistant starch are well-suited for creating easily digestible food products, particularly for infants and individuals in recovery. Similarly to waxy maize starch, millets with a high content of resistant starch, which exhibit greater paste viscosities and lower retrogradation, could be excellent choices for a range of food applications (Bangar *et al.*, 2021).

6.6.2 Thickening and Stabilizer

Food products use millet starch as a thickener and stabilizer. Most importantly, chemically modified starch opens up possibilities for application as an stabilizer, emulsifier, drying aid, thickener, dusting agent, suspending agent, clouding agent, binder, carrier agent of flavor oils, and to improve freeze-thaw stability (Shaikh *et al.*, 2020). Shaikh and colleagues demonstrated, for instance that millet starch may replace maize starch as thickener in white sauce (Shaikh *et al.*, 2020). Modified starch also displays effective performance as a dough stabilizer, while in the case of cheese bread, the final product is of finer quality with lesser acidity and lesser softness (Anjos *et al.*, 2014). In this respect, it has been underlined that making use of pre-cooked millet starch may decrease total soluble solids with the aim to reduce preparation of fermented beverages. As Hepute *et al.* (2017) show, increased stability implies that fewer suspended particles settled during storage.

Starch modified through HMT is utilized in food items like soups and gravies because the modification leads to reduced swelling capacity and solubility. Onyango and his research team investigated that mixing hydrothermally treated FMS with wheat flour enhanced the stability of the dough while baking bread (Onyango *et al.*, 2020). The bread produced achieves a particular volume, leading to a decrease in crumb rigidity and chewiness. Microwave-modified starch is recommended for delaying bread staling because of its reduced SB viscosity. Additionally, its remarkable thermal stability makes it advantageous for use in both baked and extruded products (Oyeyinka *et al.*, 2021).

Growing concerns over gluten-free options have resulted in a rise in the availability of starch-based pasta and noodles. Currently, these NS-based alternatives fall short in terms of texture, structural integrity, and sensory appeal compared to their gluten-containing counterparts. However, this limitation has been addressed through the use of modified starch in the production of noodles and pasta (Obadi & Xu, 2021). Modified starch created using chemical, physical, and enzymatic techniques greatly affects the stability, dough consistency, texture, cooking properties, and sensory characteristics of noodles and pasta (Obadi & Xu, 2021). In a study, it has been reported that incorporating modified starch into pasta items like noodles and macaroni is well-received and leads to minimal cooking loss (Ilamaran *et al.*, 2021). Here, the research conducted on the hypoglycemic properties of these foods demonstrated a low glycemic index and fewer calories, making them appropriate for managing lifestyle-related health issues. More so, more comprehensive studies are requaired on the use of modified millet starch in baked goods and noodles.

6.6.3 Finger Millet Starch in Industry

FMS act as alternative to fats for flavor encapsulation, help stabilize emulsions in drinks, provide stability to foams in marshmallows, and serve as gelling agents in gums. Further, these starches are used to augment the crispiness of fried food and in the production of tablets for medicinal delivery (Mahajan *et al.*, 2021). Finger millet undergoes several processing methods such as popping, malting, milling, decortication, and fermentation. This grain is utilized to create a range of products, including vermicelli, noodles, pasta, Indian sweets like halwa, soups, papads, and various baked goods. In addition, innovative products like soups, rolled or dried papads, and noodles are gaining popularity. A new variant of decorticated finger millet has also been introduced (Shobana *et al.*, 2013). Finger millet vermicelli and extruded snacks have a lower glycemic index than various wheat-based snacks and vermicelli, making them a preferable choice. In Kerala, India, the Muthuvan tribe refers to finger millet as "Kepa," and it is used to make a pudding called "Katty." In a study of healthy subjects, it was seen that the ready-to-cook finger millet flakes contained 2.25 g of protein, 0.13 g of minerals, and 9.67 mg of iron in each 30 g serving. These flakes were also found to have a low glycemic load with a moderate glycemic index and very low trans fats with a high fiber content. The Ready to Cook (RTC) flakes had a high acceptance score of 81.11 and hence could be a good addition to the diet (Saleem *et al.*, 2023). Finger millet is utilized in the cosmetic industry because of its high levels of antioxidants, vitamins, and minerals that promote skin health and revitalization (Kour & Nayik, 2021).

Alterations to FMS serve numerous purposes in the food sector and beyond (Fig. 6.8). Techniques like esterification enhance the favorable properties of starch and are commonly recognized in a range of food applications (Haq *et al.*, 2019). By forming a strong barrier between oil and water, octenyl succinic anhydride starch improves emulsification. In contrast, the qualities of transparency, swelling, and viscosity are enhanced by octenyl

FIGURE 6.8
Application of FMS in various sectors (Bhasin *et al.*, 2024).

succinylated and succinylated starches making them valuable in food products such as ketchup and sauces (Gao *et al.*, 2022). The process of acid hydrolysis is used to modify FMS, resulting in the production of cationic starch, starch gum candies, amphoteric starch, and paper (Bangar *et al.*, 2021). When FMS undergoes techniques such as microwave treatment, acid treatment, and hydrothermal alteration, its ability to expand is reduced but its amylose concentration is increased, thereby making it more suitable for noodle production (Sun *et al.*, 2022).

Tablet formulations incorporated FMSs in various forms, including acetylated, oxidized, pre-gelatinized, acid-treated, and native. These starches were combined with different medications such as calcium carbonate, sodium bicarbonate, sulpha imidine, chloroquine, and chloroquine phosphate (Zhu, 2014a). Pharmaceutical excipients requiring rapid disintegration and high compressive strength may be made from acetylated FMS. However, because of its acceptable solubility, high bulk and tap density, and outstanding flow properties, oxidized starch may be employed as a filler in capsule formulations (Shuaibu, 2017).

6.6.4 Food Packaging

The accessibility, biodegradability, ease of isolation, non-toxicity, affordability, flexibility, and impermeable, starch-based films make them promising for application in food packaging. Moreover, it was reported that films produced from several types of starches exhibited semi-permeability to carbon dioxide, moisture, and oxygen, while also being devoid of odor, taste, and color. The use of millet grains to produce edible and bio-degradable film is an innovative approach in the starch business. According to a research by G. Al-Hashimi *et al.* (2020) titled "Development of a Millet Starch Edible Film Containing Clove Essential Oil," adding encapsulated eugenol (EO) to starch-based films may change their characteristics, especially by making them more hydrophobic. The decrease in solubility is a result of the link between that exists between EO and starch's hydroxyl groups, which restricts these groups' capacity to interact with water molecules (G. Al-Hashimi *et al.*, 2020).

6.7 Conclusions and Future Prospects

FMS has the potential to contribute to food security and sustainable development, particularly in regions where finger millet is widely cultivated. FMS is a good source of dietary fiber, resistant starch, and minerals, making it a nutritious and potentially health-promoting ingredient. Despite its high starch content, millet is often overlooked as a starch source because of limited awareness, its starch properties are not well understood, and its starch properties are not fully exploited. However, there is growing interest in FMS due to its unique properties, nutritional benefits, and gluten-free nature, making it an excellent option for individuals with celiac disease. The gluten-free nature of FMS makes it an attractive ingredient for food manufacturers catering to the growing demand for gluten-free products. The starch has excellent functional properties, including high viscosity, stability, and emulsification, making it suitable for uses like thickening, binding, and stabilizing. Modified FMS can be used in a wide range of products, such as baked goods, sauces, and snacks. Many researchers have successfully modified and processed FMS to enhance its applications employing chemical, physical, and enzymatic modification, notably in baked goods, soups, sauces, and dairy products. Researchers and food manufacturers are exploring ways to overcome challenges and tap into the potential of FMS. Emerging food processing techniques, such as ultrasonication, HPP and other modification techniques, are being explored for modifying FMS to enhance its properties and expand its applications. Understanding and modifying FMS can provide a competitive advantage in the food industry, as it offers unique and nutritious ingredients and opens up opportunities for creating innovative, nutritious food products, contributing to its potential widespread use in the food industry and also in cosmetics, pharmaceuticals, and biodegradable packaging. FMS is not standardized, which affects its quality and consistency. Therefore, global initiatives for standardizing methods in millet flour and starch research are needed to develop new food products, increase cultivation areas, and boost farmer incomes, which would benefit both the economy and local populations dependent on millets as staple foods. Further research is needed to fully explore the potential of FMS and to overcome challenges in its production and processing.

References

Adebowale, K., Afolabi, T., & Olu-Owolabi, B. (2005). Hydrothermal treatments of finger millet (*Eleusine coracana*) starch. *Food Hydrocolloids, 19*(6), 974–983.

Alcázar-Alay, S. C., & Meireles, M. A. A. (2015). Physicochemical properties, modifications and applications of starches from different botanical sources. *Food Science and Technology, 35*, 215–236.

Amarnath, M., Muhammed, A., Antony, A. K., Malini, B., & Sunil, C. (2023). White finger millet starch: Physical modification (annealing and ultrasound), and its impact on physicochemical, functional, thermal and structural properties. *Food and Humanity, 1*, 599–606.

Anitha, S., Givens, D. I., Botha, R., Kane-Potaka, J., Sulaiman, N. L. B., Tsusaka, T. W., & Bhandari, R. K. (2021). Calcium from finger millet—A systematic review and meta-analysis on calcium retention, bone resorption, and in vitro bioavailability. *Sustainability, 13*(16), 8677.

Annor, G. A., Marcone, M., Bertoft, E., & Seetharaman, K. (2013). In vitro starch digestibility and expected glycemic index of kodo millet (*Paspalum scrobiculatum*) as affected by starch–protein–lipid interactions. *Cereal Chemistry, 90*(3), 211–217.

Annor, G. A., Marcone, M., Bertoft, E., & Seetharaman, K. (2014a). Physical and molecular characterization of millet starches. *Cereal Chemistry, 91*(3), 286–292.

Annor, G. A., Marcone, M., Bertoft, E., & Seetharaman, K. (2014b). Unit and internal chain profile of millet amylopectin. *Cereal Chemistry, 91*(1), 29–34.

Annor, G. A., Tyl, C., Marcone, M., Ragaee, S., & Marti, A. (2017). Why do millets have slower starch and protein digestibility than other cereals? *Trends in Food Science & Technology, 66*, 73–83.

Anwar, S. H., Hasni, D., Rohaya, S., Antasari, M., & Winarti, C. (2020). The role of breadfruit OSA starch and surfactant in stabilizing high-oil-load emulsions using high-pressure homogenization and low-frequency ultrasonication. *Heliyon, 6*(7), e04341.

Arns, B., Paraginski, R. T., Bartz, J., de Almeida Schiavon, R., Elias, M. C., da Rosa Zavareze, E., & Dias, A. R. G. (2014). The effects of heat–moisture treatment of rice grains before parboiling on viscosity profile and physicochemical properties. *International Journal of Food Science & Technology, 49*(8), 1939–1945.

Ashogbon, A. O. (2021). Dual modification of various starches: Synthesis, properties and applications. *Food Chemistry, 342*, 128325.

Babu, A. S., & Mohan, R. J. (2019). Influence of prior pre-treatments on molecular structure and digestibility of succinylated foxtail millet starch. *Food Chemistry, 295*, 147–155.

Balakumaran, M., Nath, K. G., Giridharan, B., Dhinesh, K., Dharunbalaji, A., Malini, B., & Sunil, C. (2023). White finger millet starch: Hydrothermal and microwave modification and its characterisation. *International Journal of Biological Macromolecules, 242*, 124619.

Balasubramanian, S., Sharma, R., Kaur, J., & Bhardwaj, N. (2011). Isolation, modification and characterization of finger millet (*Eleucine coracana*) starch. *Journal of Food Science and Engineering, 1*(5), 339.

Bangar, S. P., Ashogbon, A. O., Dhull, S. B., Thirumdas, R., Kumar, M., Hasan, M., …, & Pathem, S. (2021). Proso-millet starch: Properties, functionality, and applications. *International Journal of Biological Macromolecules, 190*, 960–968.

Bangar, S. P., Ashogbon, A. O., Whiteside, W. S., & Chaudhary, V. (2024). Lily starch—An underutilized and non-conventional source: Properties and applications. *Starch-Stärke, 76*(1–2), 2300019. https://doi.org/10.1002/star.202300019

Bangar, S. P., Balakrishnan, G., Navaf, M., & Sunooj, K. V. (2024). Recent advancements on barnyard millet starch: A sustainable alternative to conventional starch. *Starch-Stärke*, 2300232.

Bhasin, A., Maibam, B. D., & Sharma, M. (2024). Review on finger millet (*Eleusine coracana* L.): Nutritional composition, modification, its effect on physicochemical, structural, functional properties, and its applications. *Journal of Food Composition and Analysis*, 106623.

Bhatt, P., Kumar, V., Rastogi, H., Malik, M. K., Dixit, R., Garg, S., … & S. (2023). Functional and tableting properties of alkali-isolated and phosphorylated barnyard millet (*Echinochloa esculenta*) starch. *ACS Omega, 8*(33), 30294–30305.

Bojarczuk, A., Skąpska, S., Khaneghah, A. M., & Marszałek, K. (2022). Health benefits of resistant starch: A review of the literature. *Journal of Functional Foods, 93*, 105094.

Bora, P., Ragaee, S., & Marcone, M. (2019). Characterisation of several types of millets as functional food ingredients. *International Journal of Food Sciences and Nutrition, 70*(6), 714–724.

Bravo, L., Siddhuraju, P., & Saura-Calixto, F. (1998). Effect of various processing methods on the in vitro starch digestibility and resistant starch content of Indian pulses. *Journal of Agricultural and Food Chemistry, 46*(11), 4667–4674.

Brites, C. M., Santos, C. A. L., Bagulho, A. S., & Beirão-da-Costa, M. L. (2008). Effect of wheat puroindoline alleles on functional properties of starch. *European Food Research and Technology, 226*, 1205–1212.

Cai, L., & Shi, Y.-C. (2010). Structure and digestibility of crystalline short-chain amylose from debranched waxy wheat, waxy maize, and waxy potato starches. *Carbohydrate Polymers, 79*(4), 1117–1123.

Canton, H. (2021). Food and agriculture organization of the United Nations—FAO. In *The Europa Directory of International Organizations 2021* (pp. 297–305). Routledge.

Castro, L. M., Alexandre, E. M., Saraiva, J. A., & Pintado, M. (2020). Impact of high pressure on starch properties: A review. *Food Hydrocolloids, 106*, 105877.

Chandi, G., & Annor, G. (2015). Millet minor: Overview. *The World Of Food Grains*, 199–208.

Chandra, S., Singh, S., & Kumari, D. (2015). Evaluation of functional properties of composite flours and sensorial attributes of composite flour biscuits. *Journal of Food Science and Technology, 52*, 3681–3688.

Chinma, C. E., Ibrahim, P. A., Adedeji, O. E., Ezeocha, V. C., Ohuoba, E. U., Kolo, S. I., …, & Adebo, O. A. (2022). Physicochemical properties, in vitro digestibility, antioxidant activity and consumer acceptability of biscuits prepared from germinated finger millet and Bambara groundnut flour blends. *Heliyon, 8*(10), e10849.

Cornejo-Ramírez, Y. I., Martínez-Cruz, O., Del Toro-Sánchez, C. L., Wong-Corral, F. J., Borboa-Flores, J., & Cinco-Moroyoqui, F. J. (2018). The structural characteristics of starches and their functional properties. *CyTA-Journal of Food, 16*(1), 1003–1017.

Dhanushkodi, V., Hemavathy, A. T., Shenbagavalli, S., Sangeetha, S., Anitha, R., & UmaMaheshwari, T. (2023). A review on nutritional properties and health benefits of finger millet. *International Journal of Plant & Soil Science, 35*(18), 753–761.

Dharmaraj, U., Parameswara, P., Somashekar, R., & Malleshi, N. G. (2014). Effect of processing on the microstructure of finger millet by X-ray diffraction and scanning electron microscopy. *Journal of Food Science and Technology, 51*, 494–502.

Dimri, S., Aditi, Bist, Y., & Singh, S. (2023). Oxidation of starch. In *Starch: Advances in Modifications, Technologies and Applications* (pp. 55–82). Springer.

dos Anjos, L. D., Pereira, J., Couto, E. M., & Cirillo, M. A. (2014). Modified starches or stabilizers in preparation of cheese bread. *Ciência Rural, 44*(9), 1686–1691.

El Halal, S. L. M., Kringel, D. H., Zavareze, E., & Dias, d. R. (2019). Methods for extracting cereal starches from different sources: A review. *Starch-Stärke, 71*(11–12), 1900128.

Gaffa, T., Yoshimoto, Y., Hanashiro, I., Honda, O., Kawasaki, S., & Takeda, Y. (2004). Physicochemical properties and molecular structures of starches from millet (*Pennisetum typhoides*) and Sorghum (*Sorghum bicolor* L. Moench) cultivars in Nigeria. *Cereal Chemistry, 81*(2), 255–260.

G. Al-Hashimi, A., Ammar, A. B., Cacciola, F., & Lakhssassi, N. (2020). Development of a millet starch edible film containing clove essential oil. *Foods, 9*(2), 184.

Gao, W., Sui, J., Yu, B., Liu, P., & Cui, B. (2022). Effect of shell separation pretreatment on the physicochemical properties of octenyl succinic anhydride-modified starch. *International Journal of Biological Macromolecules, 221*, 1–7.

Gautam, N., Garg, S., & Yadav, S. (2021). Underutilized finger millet crop for starch extraction, characterization, and utilization in the development of flexible thin film. *Journal of Food Science and Technology, 58*, 4411–4419.

Gautam, N., Garg, S., & Yadav, S. (2022). Development of flexible and thin films from underutilized Indian finger millet (*Eleusine coracana*). *Starch-Stärke, 74*(3–4), 2100156.

Gautam, N., Garg, S., & Yadav, S. (2023). Dry heat treatment of finger millet starch for thin films—Studies on physicochemical, mechanical, and barrier properties. *Starch-Stärke*, 74(3–4), 2300097.

Geetha, K., Yankanchi, G. M., Hulamani, S., & Hiremath, N. (2020). Glycemic index of millet based food mix and its effect on pre diabetic subjects. *Journal of Food Science and Technology, 57*, 2732–2738.

Grgić, I., Ačkar, Đ, Barišić, V., Vlainić, M., Knežević, N., & Medverec Knežević, Z. (2019). Nonthermal methods for starch modification—A review. *Journal of Food Processing and Preservation, 43*(12), e14242.

Gunaratne, A., & Corke, H. (2007). Functional properties of hydroxypropylated, cross-linked, and hydroxypropylated cross-linked tuber and root starches. *Cereal Chemistry, 84*(1), 30–37.

Gunaratne, A., & Hoover, R. (2002). Effect of heat–moisture treatment on the structure and physicochemical properties of tuber and root starches. *Carbohydrate Polymers, 49*(4), 425–437.

Gupta, R., & Gaur, S. (2024). Investigating the effect of natural fermentation in modifying the physico-functional, structural and thermal characteristics of pearl and finger millet starch. *Journal of the Science of Food and Agriculture, 104*(4), 2440–2448.

Haq, F., Yu, H., Wang, L., Teng, L., Haroon, M., Khan, R. U., …, & Nazir, A. (2019). Advances in chemical modifications of starches and their applications. *Carbohydrate Research, 476*, 12–35.

Hepute, N., Embashu, W., Cheikhyoussef, A., & Nantanga, K. K. (2017). Effect of pre-cooking pearl millet (*Pennisetum glaucum*) flour on the reduction of dregs in Oshikundu. *International Science and Technology Journal of Namibia*, 091–097.

Hoover, R. (2001). Composition, molecular structure, and physicochemical properties of tuber and root starches: A review. *Carbohydrate Polymers, 45*(3), 253–267.

Hoover, R. (2010). The impact of heat-moisture treatment on molecular structures and properties of starches isolated from different botanical sources. *Critical Reviews in Food Science and Nutrition, 50*(9), 835–847.

Hu, A., Li, Y., & Zheng, J. (2019). Dual-frequency ultrasonic effect on the structure and properties of starch with different size. *LWT, 106*, 254–262.

Ilamaran, M., Sarojinibharathi, R., & Selvi, J. (2021). Development of technology for modified starch incorporated grains and pulse blended bakery and pasta products. *Journal of Applied and Natural Science, 13*(SI), 179–187.

Jhan, F., Shah, A., Gani, A., Ahmad, M., & Noor, N. (2020). Nano-reduction of starch from underutilised millets: Effect on structural, thermal, morphological and nutraceutical properties. *International Journal of Biological Macromolecules, 159*, 1113–1121.

Jia, R., Cui, C., Gao, L., Qin, Y., Ji, N., Dai, L., …, & Sun, Q. (2023). A review of starch swelling behavior: Its mechanism, determination methods, influencing factors, and influence on food quality. *Carbohydrate Polymers*, 321, 121260.

Jideani, I., Takeda, Y., & Hizukuri, S. (1996). Structures and physiochemical properties of starches from acha (*Digitaria exilis*), Iburu (*D. iburua*), and tamba (*Eleusine coracana*). *Cereal Chemistry, 73*(6), 677–685.

Kaimal, A. M., Mujumdar, A. S., & Thorat, B. N. (2021). Resistant starch from millets: Recent developments and applications in food industries. *Trends in Food Science & Technology, 111*, 563–580.

Karki, D. B., & Kharel, G. P. (2012). Effect of finger millet varieties on chemical characteristics of their malts. *African Journal of Food Science, 6*(11), 308–316.

Kaur, P., & Annapure, U. S. (2024a). Rheological and gelling properties of atmospheric pressure cold plasma treated finger millet (*Eleusine coracana*) starch. *Food Research International, 187*, 114418.

Kaur, P., & Annapure, U. S. (2024b). Understanding the atmospheric cold plasma-induced modification of finger millet (*Eleusine coracana*) starch and its related mechanisms. *International Journal of Biological Macromolecules, 268*, 131615.

Koch, K., & Jane, J. L. (2000). Morphological changes of granules of different starches by surface gelatinization with calcium chloride. *Cereal Chemistry, 77*(2), 115–120.

Kour, J., & Nayik, G. A. (2021). *Nutraceuticals and Health Care*. Academic Press.

Kumar, N., Singh, R., Upadhyay, A., & Giri, B. S. (2024). Application and functional properties of millet starch: Wet milling extraction process and different modification approaches. *Heliyon, 10*(3), e25330.

Lawal, O. S. (2009). Starch hydroxyalkylation: Physicochemical properties and enzymatic digestibility of native and hydroxypropylated finger millet (*Eleusine coracana*) starch. *Food Hydrocolloids, 23*(2), 415–425.

Li, Y., Hu, A., Wang, X., & Zheng, J. (2019). Physicochemical and in vitro digestion of millet starch: Effect of moisture content in microwave. *International Journal of Biological Macromolecules, 134*, 308–315.

Li, Y., Hu, A., Zheng, J., & Wang, X. (2019). Comparative studies on structure and physiochemical changes of millet starch under microwave and ultrasound at the same power. *International Journal of Biological Macromolecules, 141*, 76–84.

Lu, H., Zhang, J., Liu, K.-b, Wu, N., Li, Y., Zhou, K., …, & Yang, X. (2009). Earliest domestication of common millet (*Panicum miliaceum*) in East Asia extended to 10,000 years ago. *Proceedings of the National Academy of Sciences, 106*(18), 7367–7372.

Mahajan, P., Bera, M. B., Panesar, P. S., & Chauhan, A. (2021). Millet starch: A review. *International Journal of Biological Macromolecules, 180*, 61–79.

Maharajan, T., Ceasar, S. A., & Ajeesh Krishna, T. P. (2022). Finger millet (*Eleusine coracana* (L.) Gaertn): Nutritional importance and nutrient transporters. *Critical Reviews in Plant Sciences, 41*(1), 1–31.

Mehboob, S., Ali, T. M., Sheikh, M., & Hasnain, A. (2020). Effects of cross linking and/or acetylation on sorghum starch and film characteristics. *International Journal of Biological Macromolecules, 155*, 786–794.

Muñoz-Llandes, C. B., Martínez-Villaluenga, C., Palma-Rodríguez, H. M., Román-Gutiérrez, A. D., Castro-Rosas, J., & Guzmán-Ortiz, F. A. (2023). Effect of germination on starch. In *Starch: Advances in Modifications, Technologies and Applications* (pp. 457–486). Springer.

Obadi, M., & Xu, B. (2021). Review on the physicochemical properties, modifications, and applications of starches and its common modified forms used in noodle products. *Food Hydrocolloids, 112*, 106286.

Ojijo, N. K., & Shimoni, E. (2007). Influence of xanthan gum and tapioca starch on the retrogradation and gelation of finger millet (*Eleusine coracana* L. Gaertner) starch pastes. *Journal of Texture Studies, 38*(1), 100–115.

Onyango, C., Luvitaa, S. K., Unbehend, G., & Haase, N. (2020). Physico-chemical properties of flour, dough and bread from wheat and hydrothermally-treated finger millet. *Journal of Cereal Science, 93*, 102954.

Oyeyinka, S. A., Akintayo, O. A., Adebo, O. A., Kayitesi, E., & Njobeh, P. B. (2021). A review on the physicochemical properties of starches modified by microwave alone and in combination with other methods. *International Journal of Biological Macromolecules, 176*, 87–95.

Pandey, A. K., Thakur, S., & Yadav, D. K. (2023). Dry heat treatment of starch. In *Starch: Advances in Modifications, Technologies and Applications* (pp. 151–171). Springer.

Paramitasari, D., Musa, M., Putra, O. N., Suparman, S., Pramana, Y. S., Elisa, S., …, & Pudjianto, K. (2024). Hydroxypropylation for functional enhancement of sago starch: The effects of low propylene oxide concentration using response surface methodology. *Journal of Agriculture and Food Research, 15*, 100933.

Park, S. H., Na, Y., Kim, J., Kang, S. D., & Park, K.-H. (2018). Properties and applications of starch modifying enzymes for use in the baking industry. *Food Science and Biotechnology, 27*, 299–312.

Pérez, S., & Bertoft, E. (2010). The molecular structures of starch components and their contribution to the architecture of starch granules: A comprehensive review. *Starch-Stärke, 62*(8), 389–420.

Pranoto, Y. (2022). Starch and its derivatives as potential source of prebiotics. In *Probiotics, Prebiotics and Synbiotics: Technological Advancements towards Safety and Industrial Applications* (pp. 378–406). Wiley.

Pratiwi, M., Faridah, D. N., & Lioe, H. N. (2018). Structural changes to starch after acid hydrolysis, debranching, autoclaving-cooling cycles, and heat moisture treatment (HMT): A review. *Starch-Stärke, 70*(1–2), 1700028.

Premavalli, K., Jagannath, J., Majumdar, T., & Bawa, A. (2005). Studies on phase transition in finger millet starch in relation to gelatinisation. *Journal of Food Science and Technology (Mysore), 42*(4), 336–340.

Punia, S., Kumar, M., Siroha, A. K., Kennedy, J. F., Dhull, S. B., & Whiteside, W. S. (2021). Pearl Millet grain as an emerging source of starch: A review on its structure, physicochemical properties, functionalization, and industrial applications. *Carbohydrate Polymers, 260*, 117776.

Ratnayake, W., Hoover, R., Shahidi, F., Perera, C., & Jane, J. (2001). Composition, molecular structure, and physicochemical properties of starches from four field pea (Pisum sativum L.) cultivars. *Food Chemistry, 74*(2), 189–202.

Saini, S., Saxena, S., Samtiya, M., Puniya, M., & Dhewa, T. (2021). Potential of underutilized millets as Nutri-cereal: An overview. *Journal of Food Science and Technology*, 1–13.

Saleem, S., Mushtaq, N. U., Shah, W. H., Rasool, A., Hakeem, K. R., Seth, C. S., …, & Rehman, R. U. (2023). Millets as smart future food with essential phytonutrients for promoting health. *Journal of Food Composition and Analysis, 124*, 105669.

Saleh, A. S., Zhang, Q., Chen, J., & Shen, Q. (2013). Millet grains: Nutritional quality, processing, and potential health benefits. *Comprehensive Reviews in Food Science and Food Safety, 12*(3), 281–295.

Shahidi, F., & Chandrasekara, A. (2013). Millet grain phenolics and their role in disease risk reduction and health promotion: A review. *Journal of Functional Foods, 5*(2), 570–581.

Shaikh, M., Ali, T. M., & Hasnain, A. (2020). Comparative study on the application of chemically modified corn and pearl millet starches in white sauce. *Journal of Food Processing and Preservation*, *44*(4), e14393.

Sharma, R., Gautam, N., Giribabu, G., Kumar, R., & Yadav, S. (2023). Synergistic effect of plasticizer and cross-linking agent on the properties of finger millet starch-based film. *Starch-Stärke*, *75*(9–10), 2200251.

Sherin, A. J., Sunil, C., Chidanand, D., & Venkatachalapathy, N. (2024). Structural, physicochemical and functional properties of high-pressure modified white finger millet starch. *International Journal of Biological Macromolecules*, *261*, 129919.

Shobana, S., Krishnaswamy, K., Sudha, V., Malleshi, N., Anjana, R., Palaniappan, L., & Mohan, V. (2013). Finger millet (Ragi, *Eleusine coracana* L.): A review of its nutritional properties, processing, and plausible health benefits. *Advances in Food and Nutrition Research*, *69*, 1–39.

Shuaibu, K. F. B. (2017). *Evaluation of Modified Finger Millet Starch (Eleusine coracana l.) as filler/binder in Direct Compression*. Ahmadu Bello University.

Singh, H., Blennow, A., Gupta, A. D., Kaur, P., Dhillon, B., Sodhi, N. S., & Dubey, P. K. (2022). Pulsed light, pulsed electric field, and cold plasma modification of starches: Technological advancements & effects on functional properties. *Journal of Food Measurement and Characterization*, *16*(5), 4092–4109.

Singh, V., & Ali, S. (2008). Properties of starches modified by different acids. *International Journal of Food Properties*, *11*(3), 495–507.

Sinhmar, A., Pathera, A. K., Sharma, S., Nehra, M., Thory, R., & Nain, V. (2023). Impact of various modification methods on physicochemical and functional properties of starch: A review. *Starch-Stärke*, *75*(1–2), 2200117.

Sun, Q., Zhu, X., Si, F., & Xiong, L. (2015). Effect of acid hydrolysis combined with heat moisture treatment on structure and physicochemical properties of corn starch. *Journal of Food Science and Technology*, *52*, 375–382.

Sun, X., Saleh, A. S., Lu, Y., Sun, Z., Zhang, X., Ge, X., …, & Li, W. (2022). Effects of ultra-high pressure combined with cold plasma on structural, physicochemical, and digestive properties of proso millet starch. *International Journal of Biological Macromolecules*, *212*, 146–154.

Sun, X., Sun, Z., Guo, Y., Zhao, J., Zhao, J., Ge, X., …, & Yan, W. (2021). Effect of twin-xuscrew extrusion combined with cold plasma on multi-scale structure, physicochemical properties, and digestibility of potato starches. *Innovative Food Science & Emerging Technologies*, *74*, 102855.

Taylor, J. R., & Emmambux, M. N. (2008). Gluten-free foods and beverages from millets. In *Gluten-Free Cereal Products and Beverages* (p. 119). Elsevier.

Tester, R. F., Karkalas, J., & Qi, X. (2004). Starch—Composition, fine structure and architecture. *Journal of Cereal Science*, *39*(2), 151–165.

Thakur, K., Sharma, S., & Sharma, R. (2023). Morphological and functional properties of millet starches as influenced by different modification techniques: A review. *Starch-Stärke*, *75*(3–4), 2200184.

Thangaraju, S., Shankar, M., Buvaneshwaran, M., & Natarajan, V. (2022). Effect of processing on the functional potential of bioactive components. In *Bioactive Components: A Sustainable System for Good Health and Well-Being* (pp. 183–207). Springer.

Vedamanickam, R., Anandan, P., Bupesh, G., & Vasanth, S. (2020). Study of millet and non-millet diet on diabetics and associated metabolic syndrome. *Biomedicine*, *40*(1), 55–58.

Wang, L., Chen, J., Lu, S., Xiao, P., Li, C., & Yi, C. (2022). Structural characterization, physicochemical properties and in vitro digestion of finger millet–resistant starch prepared by different methods. *International Journal of Food Science & Technology, 57*(1), 416–425.

Wu, Y., Lin, Q., Cui, T., & Xiao, H. (2014). Structural and physical properties of starches isolated from six varieties of millet grown in China. *International Journal of Food Properties, 17*(10), 2344–2360.

Xiao, Y., Zheng, M., Yang, S., Li, Z., Liu, M., Yang, X., …, & Liu, J. (2021). Physicochemical properties and in vitro digestibility of proso millet starch after addition of Proanthocyanidins. *International Journal of Biological Macromolecules, 168*, 784–791.

Xing, B., Yang, X., Zou, L., Liu, J., Liang, Y., Li, M., …, & Zhang, L. (2023). Starch chain-length distributions determine cooked foxtail millet texture and starch physicochemical properties. *Carbohydrate Polymers, 320*, 121240.

Yadav, D. N., & Rajan, A. (2012). Fibres as an additive for oil reduction in deep fat fried poori. *Journal of Food Science and Technology, 49*, 767–773.

Yadav, S., Mishra, S., & Pradhan, R. C. (2021). Ultrasound-assisted hydration of finger millet (*Eleusine Coracana*) and its effects on starch isolates and antinutrients. *Ultrasonics Sonochemistry, 73*, 105542.

Yang, Z., Gu, Q., Lam, E., Tian, F., Chaieb, S., & Hemar, Y. (2016). In situ study starch gelatinization under ultra-high hydrostatic pressure using synchrotron SAXS. *Food Hydrocolloids, 56*, 58–61.

Yi, M., Tang, X., Liang, S., He, R., Huang, T., Lin, Q., & Zhang, R. (2024). Effect of microwave alone and microwave-assisted modification on the physicochemical properties of starch and its application in food. *Food Chemistry, 446*, 138841.

Yogananda, K., Ramasamy, E., Vasantha Kumar, S., & Rangappa, D. (2021). Ragi (finger millet) starch-based gel electrolytes for dye-sensitized solar cell application. *Bulletin of Materials Science, 44*, 1–6.

Zheng, M., Xiao, Y., Yang, S., Liu, H., Liu, M., Yaqoob, S., …, & Liu, J. (2020). Effects of heat–moisture, autoclaving, and microwave treatments on physicochemical properties of proso millet starch. *Food Science & Nutrition, 8*(2), 735–743.

Zheng, Y., Xu, B., Shi, P., Tian, H., Li, Y., Wang, X., …, & Liang, P. (2022). The influences of acetylation, hydroxypropylation, enzymatic hydrolysis and crosslinking on improved adsorption capacities and in vitro hypoglycemic properties of millet bran dietary fibre. *Food Chemistry, 368*, 130883.

Zhi, W., Zhou, Y., Wang, R., Wang, M., Wang, W., Hu, A., & Zheng, J. (2022). Effect of microwave treatment on the properties of starch in millet kernels. *Starch-Stärke, 74*(9–10), 2200063.

Zhu, F. (2014a). Structure, physicochemical properties, and uses of millet starch. *Food Research International, 64*, 200–211.

Zhu, F. (2014b). Structure, physicochemical properties, modifications, and uses of sorghum starch. *Comprehensive Reviews in Food Science and Food Safety, 13*(4), 597–610.

Zhu, F. (2015). Impact of ultrasound on structure, physicochemical properties, modifications, and applications of starch. *Trends in Food Science & Technology, 43*(1), 1–17.

Zhu, F. (2018). Relationships between amylopectin internal molecular structure and physicochemical properties of starch. *Trends in Food Science & Technology, 78*, 234–242.

Zhu, L., Zhang, Y., Wu, G., Qi, X., Dag, D., Kong, F., & Zhang, H. (2020). Characteristics of pasting properties and morphology changes of rice starch and flour under different heating modes. *International Journal of Biological Macromolecules, 149*, 246–255.

7

*Effect of Processing on Nutritional Profile and Functional Properties of Finger Millet (*Eleusine coracana *L.)*

Ranjit Singh, Sanjaya Karki, Himani Priya, and Simmi Ranjan Kumar

7.1 Introduction

Millets are small-seeded grasses in the Poaceae family. They are primarily grown in dry subtropical and tropical areas. These nutritious grains are abundant in carbohydrates, protein, fats, dietary fiber, minerals, flavonoids, and various vitamins (Kumar et al., 2020). Serving as a crucial staple food for more than 90 million people across Africa and Asia. In India, the estimated production of millet stands at 10.23 million metric tons, cultivated over approximately 84 million hectares of land. This accounts for approximately 41% of global millet production, making India a significant contributor to the world's supply (Rasri et al., 2024; Tagade & Sawarkar, 2023). Millets are commonly used in various traditional dishes, including porridges, idli, dosa, bread, chakli, and papads (Kumar et al., 2020; Sarita & Singh, 2016). The most widely cultivated millet species include finger, proso, and pearl millet. Nutritionally, millets provide essential nutrients such as carbohydrates (60–70%), dietary fiber (12–20%), proteins (6–19%), fats (1.5–5%), minerals (2–4%), and phytochemicals, including phenolic compounds. Moreover, millets are gluten-free, making them suitable for individuals with celiac disease, and they can help lower blood sugar levels, making them beneficial for managing type II diabetes (Kaur et al., 2023; Kumar et al., 2023).

Finger millet (*Eleusine coracana* L.) is an exceptional but often overlooked cereal crop predominantly flourishing in the dry regions of Africa and Asia. Ranking fourth most crucial millet worldwide, it features a robust nutritional profile and is rich in essential nutrients, including iron, calcium, amino acids, and vitamins (Abioye et al., 2022). This versatile grain is known by various names across different cultures: it is called "bulo," "ragi," and "wimbi" in Uganda; "tamba," "pwana," and "sarga" in Nigeria; "dagussa" in Ethiopia; "ragi" and "mandua" in India; "wimbe" in East Africa; and "rapoko" in South

DOI: 10.1201/9781003535119-7

Africa. In English, it is commonly referred to as African millet, birdsfoot, or coracana (Rasri et al., 2024). The annual finger millet production is approximately 4.5 million tons, with around 2.5 million tons produced in Africa and 1.2 million tons in India. Besides Africa and India, finger millet is also cultivated in Nepal, Sri Lanka, and the Himalayan regions of India, as well as in China, Japan, South Carolina, and Taiwan. Approximately 55–60% of the global finger millet is sourced from sub-Saharan Africa (Abah et al., 2020; Abioye et al., 2022; Antony Ceasar et al., 2018). Due to its adaptability and exceptional nutritional benefits, finger millet is increasingly being acknowledged as a promising crop and a viable substitute for traditional cereal grains like maize, rice, and wheat. The finger millet grains offer numerous health benefits resulting from their consumption are extensive and varied antioxidant, hypocholesterolemic, hypoglycemic, antimicrobial, antilithiatic, anticataract, antiulcerative, anticancer, and cardioprotective properties. Consequently, finger millet is ideal for individuals suffering from malnutrition, celiac disease, and osteoporosis (Bhasin et al., 2024; Jayawardana et al., 2022). Moreover, finger millet can be utilized as a sustainable source for producing resistant starch (RS), which remains undigested in the small intestine and is instead fermented by gut bacteria. This fermentation process produces beneficial short-chain fatty acids, including acetate, butyrate, and propionate, which promote gut health and contribute to improved digestive function. Like dietary fiber, RS supports gut health, helps regulate blood sugar levels, and aids in weight management, making it an excellent ingredient for creating functional food products (Bhasin et al., 2024; Singh et al., 2021).

Because of their dense nutrient composition and tough texture, millets cannot be eaten raw or uncooked. Processing is essential to improve the bioavailability and bioaccessibility of their nutrients, as well as to enhance digestibility (Shukla et al., 2024). Various techniques are employed towards enhance grains' nutritional and functional quality while reducing anti-nutrients. In the case of millets, techniques such as sprouting, roasting, fermentation, and popping can improve their nutritional and functional properties (Rathore et al., 2019). Over time, numerous food processing techniques have been developed to enhance the final product's appeal in terms of texture and flavor while ensuring consistency (Samtiya et al., 2021). Millets can be processed using both conventional and novel methods. Traditional techniques include cooking, popping, steaming, parboiling, and fermentation, while novel methods encompass thermal processes such as ohmic heating, microwave, infrared, and radiofrequency treatments, coupled with non-thermal techniques such as ultrasonication, irradiation, electric fields, ozonation, and cold plasma (Joshi et al., 2023; Shukla et al., 2024). These processing techniques are known to enhance nutrient bioavailability and bioaccessibility in millets, reduce dry mass, and improve functional properties such as water and oil holding capacity. Additionally, these methods enhance enzymatic activity and improve the grain's microstructure, further contributing to better nutritional and functional properties (Gowda et al., 2022). This chapter focuses on the impact of different processing techniques on the nutritional profile and functional properties of finger millet. It provides valuable insights into how these methods influence the grain's nutritional and functional characteristics, along with its applications in various food products, including bread, pasta, muffins, porridge, and dairy products.

7.2 Nutritional Aspects of Finger Millet

Finger millet, or ragi, is a nutrient-dense cereal that stands out among other millets and cereal grains due to its unique nutritional profile (Tables 7.1 and 7.2). It is rich in calcium, iron, and dietary fiber, making it a valuable addition to a balanced diet. Compared to other millets like pearl millet, proso millet, foxtail millet, and barnyard millet, finger millet offers higher calcium content but varies in protein and fiber levels. For example, pearl millet has a higher protein content but lower calcium, while barnyard millet surpasses finger millet in fiber content but contains less calcium (Jayawardana et al., 2019). These characteristics make finger millet suitable for therapeutic diets, especially for populations with high calcium and iron needs. Its unique nutritional profile makes it stand out among other cereals and millets like rice and wheat, contributing to its potential as a reliable dietary staple, especially in regions prone to nutrient deficiencies

TABLE 7.1

Comparative Nutritional Composition of Finger Millet, Rice, and Wheat (per 100g)

Nutrient	Finger Millet	Rice	Wheat	References
Carbohydrates (%)	72–75	80.4	71.2	Dharmaraj and Malleshi (2011)
Protein (%)	5.6–12.7	6.8	11.8	Mashaghala et al. (2023)
Fat (%)	1.3–1.8	0.5	1.5	Sridhar (1994)
Dietary Fiber (%)	11.5–18	0.4	2.7	Vinutha et al. (2024), Devi et al. (2014)
Calcium(mg)	344	10	30	Jayawardana et al. (2019)
Iron(mg)	3.9–6	0.7	3.1	Dharmaraj and Malleshi (2011), Jayawardana et al. (2019)
Magnesium (mg)	134	19	136	Jayawardana et al. (2019)
Potassium (mg)	402–412	70	340	Abioye et al. (2022)
Phosphorus (mg)	283–341	43	350	Abioye et al. (2022), Vinutha et al. (2024)

TABLE 7.2
The Composition of Millets in Terms of Amino Acids per 100 Grams of Protein

Amino acid	Finger Millet	Rice	Wheat	References
Leucine (g)	0.7	0.55	0.92	Chandra et al. (2016)
Valine(g)	0.48	0.40	0.6	Shobana et al. (2009)
Isoleucine (g)	0.40	0.55	0.53	Abioye et al. (2022)
Phenyl (g)	0.31	0.35	0.68	Vinutha et al. (2024)
Threonine (g)	0.24	0.24	0.37	Bhasin et al. (2024)
Lysine (g)	0.22	0.24	0.27	Chandra et al. (2016), Mashaghala et al. (2023), Shobana et al. (2009)
Tyrptophan (g)	0.22	0.22	0.36	Abioye et al. (2022)
Methionine (g)	0.21	0.16	0.22	Chandra et al. (2016), Bhasin et al. (2024)
Histidine (g)	0.13	0.16	0.32	Vinutha et al. (2024)

7.2.1 Carbohydrate

Finger millet is a rich source of carbohydrates, consisting of starch (43–65.5%), dietary fiber (11.5–18%), and free sugars (1.04%). The grain has substantial fiber content due to its 20–30% non-starch polysaccharides carbohydrate composition (Tripathi et al., 2022). The total starch content in finger millet ranges from 65.5% to 43%, with varietal differences, and it comprises two major fractions: amylose and amylopectin (Shobana et al., 2013). Finger millet starch has a lower amylose concentration (16%) compared to other millets, including sorghum (24.0%), pearl millet (21.0%), proso millet (28.2%), foxtail millet (17.5%), and kodo millet (24.0%) (Shobana et al., 2013). In addition to starch, finger millet includes pentosans (6.2–7.2%), cellulose (1.4–1.8%), and lignins (0.04–0.6%) (Wankhede et al., 1990). These non-starch polysaccharides provide it a higher dietary fiber content than brown rice, polished rice, and millet species as foxtail, tiny, kodo, and barnyard millet (Mashaghala et al., 2023; Patel et al., 2022). Finger millet's 11.5% dietary fiber, which comprises soluble and insoluble fractions, has digestive benefits and lowers blood sugar by reducing carbohydrate digestion and absorption. It also contains arabinose, mannose, xylose, glucose, fructose, and ribose, as well as monosaccharides such glucose, sucrose, maltose, fructose, and ribose (Dharmaraj & Malleshi, 2011). These sugars affect the grain's carbohydrate makeup and appropriateness for porridge, breads, and fermented meals.

7.2.2 Protein

Finger millet offers a protein content ranging from 5.6% to 12.70%, with 71% of its protein being digestible (Patil et al., 2023). It has a well-balanced amino acid profile, containing both essential amino acids like valine, threonine, and lysine and non-essential amino acids like proline, leucine, and phenylalanine (Anal et al., 2024). Lysine and methionine are among the most abundant essential amino acids in finger millet, which are often lacking in other starchy cereals like rice and wheat (Abioye et al., 2022). The presence of tryptophan, an essential amino acid often absent in cereals, further boosts the nutritional value of finger millet. In terms of essential amino acid composition, finger millet provides isoleucine (4.4 g), leucine (9.5 g), methionine (3.1 g), and phenylalanine (5.2 g) per 100 g, levels comparable to or exceeding those found in other cereal grains like wheat and rice (Patil et al., 2023).

7.2.3 Fat

The fat content in finger millet is relatively low, ranging between 1.2% and 1.8%, contributing to its long shelf life compared to other cereals like pearl millet (Tonapi et al., 2022). The lipid profile is predominantly composed of neutral lipids (70–72%), along with glycolipids (10–12%) and phospholipids (5–6%) (Sridhar, 1994). The primary fatty acids include oleic acid (46–62%), palmitic acid (20–35%), and linoleic acid (8–27%), making it rich in unsaturated fats that are beneficial for cardiovascular health. Saturated fatty acids make up only 25.6% of the total fat content, while unsaturated fats account for 74.4% (Sridhar, 1994).

7.2.4 Dietary Fiber and Polyphenols

Finger millet is known for its high dietary fiber content (11.5%), which surpasses that of many other cereals such as rice and wheat (0.4% and 2.7%, respectively) (Annison & Topping, 1994). The dietary fiber contributes to improved digestive health, helping increase fecal bulk, lower blood cholesterol levels, and reduce the risk of colon cancer (Devi et al., 2014). The polyphenols in finger millet, especially ferulic acid (64–96%) and p-coumaric acid (50–99%) are concentrated in the outer layers of the grain, such as the aleurone layer and testa. These polyphenols exhibit strong antioxidant properties, helping to reduce oxidative stress and inflammation, making finger millet particularly beneficial for managing metabolic diseases (Patil et al., 2023).

7.2.5 Mineral and Vitamin Content

Finger millet is an exceptional source of calcium, containing 344 mg/100 g, which is significantly higher than other cereals like rice (10 mg) and wheat (30 mg) (Vinutha et al., 2024). This makes finger millet an ideal dietary supplement for promoting bone health, particularly in children, adolescents, and the elderly. Other essential minerals include iron (3.9 mg/100 g), phosphorus (283 mg/100 g), magnesium (134 mg/100 g), and potassium (408 mg/100 g), all of which contribute to vital physiological processes such as bone development, muscle function, and blood pressure regulation (Patil et al., 2023). Finger millet is also a source of B vitamins such as niacin, thiamine, riboflavin, and folic acid, although these vitamins are present in lower amounts compared to other cereal grains. Its polyphenol and tannin content, particularly in red cultivars, has been found to affect the bioavailability of minerals and proteins, but processing techniques like malting can help improve its digestibility and nutrient absorption (Tharanathan & Mahadevamma, 2003).

7.2.6 Essential Amino Acids and Micronutrients

Finger millet offers a good balance of essential amino acids, which are vital for the body's growth and repair processes. It contains isoleucine (4.4 g), leucine (9.5 g), methionine (3.1 g), and phenylalanine (5.2 g) per 100 g, which are comparable or higher than those found in rice and wheat (Vinutha et al., 2024). It also contains notable amounts of zinc, manganese, molybdenum, and copper, which are crucial for enzymatic reactions, bone development, and immune function.

7.3 Effect of Processing on the Nutritional Profile of Finger Millet

7.3.1 Conventional Methods

7.3.1.1 Milling and Dehulling

Grain milling is a fundamental process in food preparation that has been practiced for centuries, playing a vital role in transforming dried and soaked cereal grains into flour. Traditionally, this process involves the use of wooden or stone mortars and pestles, which are tools that have been used by various cultures around the world. This process typically includes several key steps: sorting, cleaning, hulling, branning, and kilning. While traditional methods remain in use, modern milling techniques have diversified to include various equipment, such as conventional stone mills, burr mills (steel or emery type), hammer mills, and ball mills (Ramashia et al., 2019). Dehulling removes approximately 12–30% of the outer husk, bran, and germ of grains, minimizing the loss of important proteins and amino acids like histidine, lysine, and arginine. Since most antinutrients that inhibit protein absorption are found in the hull of millets, dehulling enhances protein digestibility (Nanje Gowda et al., 2022).

Krishnan et al. (2012) examined the impacts of decortication, popping, and malting on the bioavailability of calcium, iron, and zinc in finger millet. Their findings revealed that dehulling reduced calcium, iron, and zinc content by 40%, 50%, and 20%, respectively (per 100 g). However, the bioaccessibility of these minerals increased by 15%, 26%, and 24%, respectively. The millet seed coat contains high levels of phytic acid, which limits mineral absorption; thus, its removal enhances bioaccessibility. In another study on processing effects on millet antioxidant activity, dehulling decreased total phenolic content (TPC) and DPPH radical scavenging activity in both the Ravi and local finger millet varieties. Specifically, TPC decreased by 21.13% in the Ravi variety and by 12.10% in the local variety, indicating that phenolic compounds are primarily concentrated in the outer layers of the grain (Chandrasekara et al., 2012).

7.3.1.2 Roasting

Roasting is a dry-heat treatment that exposes cereal grains to high temperatures, which reduces moisture content, enhances flavor and aroma, and significantly alters the nutritional profile, anti-nutritional factors, and functional properties of millet. This process decreases anti-nutritional components such as saponins, alkaloids, trypsin inhibitors, and glycosides while improving shelf life (Ramashia et al., 2019). Dhliwayo et al. (2023) observed significant reductions in fiber, ash, tannin, phytic acid, and oxalic acid after germination, followed by roasting. The combined effects of germination and roasting considerably reduce anti-nutritional factors, as these compounds leach out during germination. A study by George et al. (2023) found that roasting at 110°C for 5 minutes significantly decreased tannins, phytates, phenols, and oxalate levels. For example, roasting reduced tannins in the IE4115 and IE3779 finger millet varieties by 6.4% and 22.75%, respectively, while phytate levels decreased by 11.14% and 19.26%. Singh et al. (2018) found that roasting affected the physiochemical and proximate composition of finger millet, resulting in a 25.02% reduction in moisture, 16% in protein, 13.64% in fat, and 3.46% in antioxidant activity, while carbohydrate and ash content increased. The decrease in protein is attributed to thermal degradation. Iron content increased by 13.33%, and calcium content rose by 1.17%. TPC decreased by 28.94%, likely due to the degradation of phenolics during heat treatment or their leaching into the endosperm, forming complexes with macromolecules that reduce their extractability.

Navyashree et al. (2022) reported that processing white finger millet (KMR-340) resulted in a 4.82% increase in carbohydrate content, while protein content decreased from 6.52% to 5.18% after roasting at 120°C for 5 minutes. This reduction is likely due to thermal degradation and the decrease in specific organic molecules containing nitrogen that can evaporate at room temperature (Kumar et al., 2020). Additionally, minerals such as calcium (Ca), magnesium (Mg), and phosphorus (P) decreased significantly, while iron (Fe) levels increased, possibly due to reduced phytic acid content after thermal treatment and iron migration from the roasting pan. Aisoni et al. (2018) examined how processing influences the physicochemical as well as proximate composition, noting that roasting for 10–15 minutes resulted in lower proximate values compared to fermented millet, with protein content decreasing from 6.46% in fermented millet to 6.53% in roasted millet. Chandrasekara et al. (2022) examined various processing methods for the antioxidant activities of value-added products such as porridges and pittu made from finger millet. Their analysis revealed a 10.26% decrease in protein content and a 5.29% reduction in fiber after roasting, while fat and ash content increased by 36.34% and 2.57%, respectively. TPC increased from 24.71 μmol ferulic acid equivalent/g dry matter to 27.93 μmol after 5 minutes of roasting, and total flavonoid content (TFC) rose from 9.63 μmol catechin equivalent/g dry matter to 11.25 μmol. These increases in TPC and TFC are likely due to the degradation and liberation of phenolic compounds during roasting (Pradeep & Sreerama, 2015). Roasting also reduces moisture content significantly, enhancing the shelf life of the product.

7.3.1.3 Germination

Germination is one of the simplest and most common traditional processing methods used to modify the biochemical and sensory properties of cereals which involves the absorption of water, that helps solubilize anti-nutrients, thereby reducing their levels (Table 7.3). The processing conditions and duration of germination influence the extent to which the concentration of anti-nutrients in cereal grains decreases during germination. Yenasew and Urga (2022) explored the impact of germination on the physicochemical characteristics of various finger millet varieties. They reported increased protein content with longer germination times, attributed to enhanced protease activity. For instance, after 48 hours of germination, the protein content of the Tessema variety rose by 21.60% compared to ungerminated millet. Similarly, a study by Azeez et al. (2022) reported a significant increase in protein content in brown finger millet after 48 hours of germination, during which total starch content decreased by 6.09%. The authors also noted a substantial reduction in anti-nutritional factors, such as phytic acid and oxalate, and significant improvements in total phenolic, flavonoid, and antioxidant activity. These findings indicate that germination effectively reduces anti-nutritional factors in millet, thereby enhancing its nutritional quality. Sharma and Garg (2023) reported increased protein content after 48 hours of germination and significant improvements in the mineral content of germinated finger millet flour, with iron levels rising from 12.39 mg/100 g to 40.12 mg/100 g. This improvement may be linked to increased phytase activity, which hydrolyzes the bonds between enzymes, minerals, and proteins, releasing the minerals and making them more bioavailable (Yenasew & Urga, 2022).

However, Kumar et al. (2021), in their study on germination time in finger millet for functional food applications, found a reduction in protein content from 6.04% to 3.41% after 96 hours of germination. This decrease is likely due to the outpacing of protein synthesis by proteolysis and the migration of nitrogenous matter to the growing embryo. The author also reported a rise in

TABLE 7.3

Transformation of Nutrients and Anti-Nutrients during Germination in the Finger Millet

Germination conditions	Objective	Results	References
Time: 48 Temperature: 23.2°C ± 2.6°C RH: 37.91± 8.17	Effect of germination	increase in protein and fiber content, with reduced ash and fat content.	Yenasew and Urga (2022)
Time: 12–96 hours	Effect of soaking and germination on the quality attributes	Significant increase in sugar, crude fiber, fat, and antioxidant activity. Significant reduction in protein content after 96 hours of germination. Tannin content was decreased compared to the ungerminated one.	Kumar et al. (2021)
Time 48 hours Temperature 25°C	Effect of germination	Significant improvement in protein content, fiber content, total phenolic and flavonoid content, and antioxidant activity. Significant reduction of phytic acid and oxalate. Improvement in Ca, Fe, Mg, P, and Zn content.	Azeez et al. (2022)
Time: 48 hours Temperature: 35°C	Effect of germination	Significant improvement in protein, ash, and fiber quantity, Reduction in fat and carbohydrates. Significant improvement in mineral content, Zn. Ca. Mg. K and Fe.	Sharma and Garg (2023)
Time: 48 hours Temperature: 20°C–23°C	Effect of germination and roasting	Considerable increase in fat, protein ccarbohydrate content, and mineral content. Reduction in fiber, ash, tannin, and oxalate	Dhliwayo et al. (2023)
Time: 24, 48, 72 hours Temperature: 22°C, 26°C, 30°C	Effect of germination	Higher processing temperatures increased TPC, but prolonged germination reduced it. Higher processing temperature lowered tannin content, with further decreases during extended germination.	Bhasin et al. (2024)
Time: 48 hours, humid atmosphere	Effect of germination and popping	Significant improvement in protein content, fat content, and fiber content. Decrease in ash and carbohydrate content. Improvement in Ca, Fe, Vit. C. Decrease in tannin, phytate, and oxalic acid.	Chauhan and Sarita (2018)

reducing sugar content from 0.86% to 10.54% and a decline in starch content after 96 hours of germination. Furthermore, the study observed an initial decrease in total phenol and tannin content, which later rose gradually after prolonged germination, due to this phenomenon leaching of phenols and tannins into water at the early stages. At the same time, their concentration increases with an increase in the total percentage of seed coat with extended germination time.

7.3.1.4 Fermentation

Fermentation is an ancient food preservation method commonly used for cereals like millet. It enhances nutritional value by increasing nutrient bioavailability and reducing anti-nutritional factors. This process also improves texture and flavor, extends shelf life, and boosts antioxidant properties, making millets more nutritious and health-promoting. Jan et al. (2022) analyzed the impact of different fermentation techniques on the nutritional profile, bioactive compounds, and anti-nutritional factors of finger millet. Their findings revealed that fermentation significantly increased the TPC and antioxidant activity. Additionally, crude protein increased dramatically from 9.10% to 12.70% after 36 hours of fermentation by using *Saccharomyces cerevisiae* while using *Lactobacillus brevis* increased to 13.13% after 36 hours of fermentation. This is likely due to the degradation of complex proteins into smaller peptides and amino acids. Moreover, tannin content was reduced by approximately 53% in both cases. Meanwhile, minerals such as Cu, Fe, Zn, and Mn increased after fermentation. Antony et al. (1996) examined primary nutrients in finger millet, observing a rise in protein from 8.1% to 8.3% after fermentation and an enhancement in protein extractability from 59.9% to 76.2%. This suggests that microbial protease activity during fermentation breaks down tannins and phytates that bind to protein, enhancing its solubility. Gabaza et al. (2016) evaluated the impact of fermentation on the phenolic content of finger millet. They reported reductions in TPC by 6–31% and total condensed tannin content by 32–52%. The hydrolysis that occurs during fermentation, driven by enzymatic action on tannin-protein and tannin-iron complexes, may lead to improved protein digestibility and enhanced bioavailability of trace minerals (Onweluzo & Nwabugwu, 2009). Abioye et al. (2022) studied natural fermentation on finger millet's nutritional and anti-nutritional composition. After 72 hours, protein content increased by 36.99%, while crude lipid and fiber levels decreased by approximately 48%. Significant increases were noted in mineral content, including phosphorus, potassium, calcium, sodium, and iron, while anti-nutritional factors decreased markedly. For instance, trypsin activity dropped by 56.16%, oxalate by 44.97%, tannin by 52.41%, and phytate by 77.06%. Additionally, Makokha et al. (2002) reported a 54.3% decrease in phytic acid content after 72 hours of fermentation and a 72.3% decrease after 96 hours. Mutshinyani et al. (2020) reported an increase in protein content in finger millet after fermentation time of 96 hours from 7.1 g/100 g of flour to 9.4 g/100 g of flour for light brown finger millet, while the fat content decreased by 22.55.%. All literature suggests that fermentation can be one of the best methods to enhance the nutritional profile of finger millet while reducing the anti-nutritional factor significantly.

7.3.2 Novel and Non-Conventional Methods

7.3.2.1 Microwave Treatment

Microwave processing, operating at frequencies between 300 MHz and 300 GHz, provides rapid and uniform heating, faster drying rates, reduced processing times, precise control, and low operational costs through direct molecular interaction with electromagnetic fields (Anal et al., 2024). This process has been selectively used to inactivate lipase in grain flours, thereby extending their shelf life (Sruthi & Rao, 2021). For instance, Ananthu et al. (2023) examined the functional, pasting, and storage stability of white finger millet flour after microwave heating at 720, 840, and 900 W for 2–6 minutes. During 15 days of accelerated storage, higher power levels and longer treatment times dramatically lowered flour's peak viscosity and free fatty acid values. Hydrothermal and microwave processing were investigated for white finger millet starch modification (Balakumaran et al., 2023). Microwave processing (510 W, 10 minutes) lowered starch pH from 6.88 to 6.35 and increased least gelation capacity from 13.63% to 16.41%.

Hithamani and Srinivasan (2014) analyzed how home-based techniques influence the polyphenol profile and bioaccessibility of both finger and pearl types of millets. Application of microwave heating (300, 450, and 600 W) for 3 minutes significantly reduced the bioaccessibility percentage of polyphenols and flavonoids, whereas 300 W treatment had a significant increase in tannin content from 5.93 mg/g to 7.85 mg/g. At 600 W treatment, protocatechuic acid was not determined, while the vanillic acid was found to be 28.37 µg/g. Furthermore, Singh et al. (2017) evaluated the effect of processing on the proximate composition, phenolic content, and anti-nutritional factors of millet. This study indicated a significant increase in protein content of finger millet from 10.07% to 14.48% and ash from 1.89 to 3.09, whereas fat content and fiber content decreased by 28.57% and 10.48%, respectively, after microwave treatment (900 W, 40–100 seconds). The study also reported a significant reduction in phytate and tannin by 58.14% and 45.18%, respectively. It suggests that microwaves can be helpful to in reducing anti-nutritional factors and enhancing the nutritional profile of millets.

7.3.2.2 Ultrasonic Treatment

Ultrasound (US) is novel non-thermal processing that enhances food quality and safety by using high-energy waves (20–100 kHz) to create acoustic cavitation, which improves heat and mass transfer through micro-streaming during processing (Anal et al., 2024). Finger millet contains anti-nutrients like tannins and phytates, which must be removed for optimal health benefits. Yadav et al. (2021) investigated ultrasound-assisted hydration on finger millet starch isolates and anti-nutrients. Their findings revealed that under optimal conditions of 66% amplitude, a 26-minute processing time, and a 1:3 grain-to-water ratio, phytate content was reduced by 66.98% and tannin content by 62.83%. Additionally, water binding capacity increased, and starch solubility improved by approximately 41%.

Amarnath et al. (2023) modified white finger millet starch through ultrasound treatment, observing a notable decrease in the amount of fat and protein., while amylose content increased from 16.23% to 25.41% under 33 kHz, 30 minutes of processing at 50°C. Through ultrasound treatment, modified white finger millet starch showed substantially decreased protein and fat content (Yadav et al., 2021).

Balasubramaniam et al. (2019) examined the impact of enzyme pretreatment on the ultrasound-assisted extraction of polyphenols from the seed coat of finger millet. Under optimal ultrasound conditions (250 W, 55°C, 25 minutes, 50% ethanol), the total yield of flavonoids and tannins significantly surpassed that of the conventional heat reflux method. In contrast to the increase in tannin from 725 to 850 mg/100 g, the flavonoid increased by 142 mg/100 g, reaching 492 mg/100 g, while antioxidant activity rose by approximately 7.69%. This enhanced radical scavenging activity is likely due to ultrasound-induced cavitation releasing bound phenolic compounds.

However, Dubey and Tripathy (2024) reported a maximum reduction of tannin and phytate content by 73% and 71%, respectively, under conditions of 80% amplitude, 20 minutes of processing, and a frequency of 20 kHz. This reduction can be attributed to acoustic cavitation, which disrupts the surface of the millet, enhancing solvent interactions and promoting the leaching of phytates and tannins.

7.3.2.3 Ohmic Treatment

Ohmic heating, also called electrical resistance or Joule heating, is an innovative method that quickly and evenly heats food by passing an electric current through it. This technique minimizes sensory changes, off-flavors, nutritional losses, and degradation of bioactive compounds (Singh et al., 2024). Although no research is currently available on the processing of finger millet using ohmic heating, studies on pearl millet have been conducted. Dias-Martins et al. (2019) performed a study that compared the effects of ohmic heating on whole and decorticated pearl millet with those of open-pan cooking. The findings indicated a notable decrease in cooking yield, with ohmic-heated whole millet decreasing by

30%. Additionally, the hardness of ohmic-heated samples was more significant than those cooked in an open pan. The study also reported a non-significant decrease in the water solubility and water absorption indices, potentially due to the more uniform heating provided by ohmic heating, which prevents grain rupture.

7.3.2.4 Extrusion

Extrusion technology converts ingredients into value-added products through shear force, high pressure, and heat, effectively modifying starch properties (Rathore et al., 2019). This process involves feed transport, mixing, and forming, enabling the mass production of nutrient-rich, affordable, and shelf-stable millet-based foods. It also allows for favorable biochemical transformations such as protein denaturation, starch gelatinization, and the inactivation of microbes, enzymes, and anti-nutritional components. Patil et al. (2016) studied the impact of extrusion on the bioactive compound of finger millet, finding a significant reduction in TPC, TFC, and antioxidant activity. Under processing conditions with a feeder speed (20 rpm), moisture content (10%), screw speed (200 rpm), and die temperature (150°C), TPC decreased by 62.68% and TFC by 47.88%. Additionally, DPPH scavenging activity dropped from 60.69 μmol TE/g dw to 27.24 μmol TE/g dw under the same conditions. In another study, Patil et al. (2016) incorporated extruded finger millet flour (20 g/100 g) into composite bread, which improved extensibility, specific volume, and loaf height while reducing firmness. This flour can effectively replace wheat flour, enhancing phenolic content and antioxidant activity with minimal impact on bread quality when used in moderation, making it a viable option for producing softer, more cohesive, functional bread.

Nishani et al. (2017) examined the extrusion processing parameters and their findings indicated that the sectional expansion index decreased as amylose content increased, likely due to higher viscosity during the cooling of extrudates, which restricted pore formation by impeding water evaporation. Additionally, higher temperatures and screw speeds increased the water solubility index, likely because more excellent mechanical energy transfer during extrusion enhanced gelatinization. The study also reported that RS content increased with higher temperatures. Liu et al. (2020) studied how extrusion conditions affect the physicochemical properties of finger millet. Results showed that the total starch content in extruded cereal products ranged from 61.63% to 74.87%, with lower moisture-reducing starch concentration due to shear degradation. Additionally, in vitro starch digestibility (IVSD) varied widely: rapidly digestible starch (RDS) from 79.5% to 90.07%, slowly digestible starch (SDS) from 8.07% to 14.1%, and RS from 1.03% to 7.0%. Moreover, higher temperatures increased starch gelatinization and digestibility, while screw speed had no significant effect on starch digestibility.

7.4 Effect of Processing on the Functional Properties of Finger Millet

7.4.1 Pasting and Rheological Properties

The effects of processing, such as germination, fermentation, heat treatment, milling, hydrothermal treatment, and extrusion, alter pasting and rheological properties of finger millet. Germination reduces peak and final viscosities due to increased enzymatic activity, while fermentation lowers the pasting temperature and peak viscosity due to microbial enzyme activity, leading to partial starch hydrolysis. Heat treatments like roasting can enhance peak viscosity by promoting starch gelatinization, while milling affects particle size and influences viscosity and rheological behavior. Finer particles often result in higher viscosity and more significant shear-thinning behavior. Hydrothermal treatments, such as parboiling, improve viscosity and gel strength through starch gelatinization. Extrusion tends to reduce viscosity and shear-thinning behavior, breaking down starch and denaturing proteins (Panghal et al., 2019). Yenasew and Urga (2022) observed that germination improves the physicochemical properties, resulting in enhanced protein content and reduced peak viscosity, with the viscosity decreasing from 1800 cP to around 1580 cP as germination progresses. The pH value of germinated finger millet also dropped from 6.43 to 5.97 after 72 hours due to the breakdown of starch into simpler sugars by enzymatic activity. Malting resulted in viscosity reductions due to starch hydrolysis, with breakdown viscosities decreasing significantly from 502 cP to 456 cP. These enzymatic changes contribute to better digestibility and functional properties in the final product

7.4.2 Texture and Structure

Finger millet has unique textural and structural properties that improve its nutritional value and make it a versatile ingredient in many dishes. Patil et al. (2016) found that gelatinized starch, aggregated protein, and degraded fiber in extruded finger millet reduced composite bread firmness (3.81–10.68 N) concerning unextruded flour (4.10–18.23 N). Since water absorption increased, extensibility and stickiness increased. Due to gluten matrix dilution and higher fiber content, unextruded finger millet flour had 85.15 N with 30% addition, compared to 65.07 N for extruded flour. In another study, Dharmaraj et al. (2015) examined finger millet functional characteristics and grain morphology following decortication after hydrothermal processing. After hydrothermal treatment (steeping for 10 hours at 30°C, then steaming in an autoclave), the grain hardness increased significantly from 37.8 N to 234.9 N but decreased to 161.3 N after decortication. The hardening of the grain presumably resulted from the voids in the endosperm being occupied by expanded starch granules during steaming. Additionally, protein bodies disintegrated and bound with starch while the cell walls compacted, leading to the overall hardening of the grain. Amarnath et al. (2023) modified white millet starch using annealing and ultrasonication. Ultrasonication reduced cohesiveness by 2.53%, gumminess by 33.33%, and hardness by 31.58%, while increasing adhesiveness from 10.76 to 24.51 kg.s and chewiness from 0.14 to 0.23. On the other hand, Annealing increased adhesiveness by 278.49% and hardness by 63.16% but decreased cohesiveness by 7.59%. These changes resulted from starch degradation during ultrasonication, which affected gel formation, while annealing led to starch rearrangement, enhancing gel hardness.

Various researchers have also investigated the FTIR and X-ray diffraction (XRD) patterns of finger millet. Gowthamraj et al. (2021) studied the fermentation effect on protein profile and microstructural properties. FTIR analysis showed maximum absorption of amide I (1690–1620 cm^{-1}) in the control sample, as well as in samples treated with sodium chloride and sample subjected to 20 hours of fermentation, indicating a predominantly alpha-helix structure. This structure was altered when exposed to lower pH. Similarly, the study on the processing effect on white finger millet using FTIR analysis revealed significant differences in functional groups and molecular interactions among native flour (Nf), roasted flour (Rf), and germinated flour (Gf) (Navyashree et al., 2022). The broad peaks between 3500 and 3000 cm^{-1}, attributed to O-H bond stretching, narrowed in Rf and Gf, indicating hydrogen bond formation and de-hydroxylation in Rf due to roasting (Sharanagat et al., 2019; Sun et al., 2014). The shift towards lower peaks signified improved bond strength. Variations in peak intensity (amide I region) in Rf were linked to Maillard reactions, while changes in Gf were due to starch depolymerization and protein modifications during germination (Kaur & Gill, 2021; Sharanagat et al., 2019). Arzoo et al. (2024) examined the structural properties of finger millet. FTIR analysis revealed consistent peaks across treatments, with notable intensity variations. Peaks at 3300–3200 cm^{-1}, associated with O-H bond stretching, increased in parboiled, germinated, and popped samples, indicating the presence of a free O-H group. Peaks in 1730–1750 cm^{-1} reflected increased fat content, particularly in popped millet due to added mustard oil. Similarly, Amarnath et al. (2023) reported the FTIR spectra of native (NS) and modified WFM starch showed no new peaks, indicating that both annealing and ultrasound treatments were purely physical modifications. However, transmittance and peak intensity varied across modified starches. For example, the peak at 997 cm^{-1} (α-1,4 glycosidic linkages) had higher intensity in annealed starch (AS) but was lower in ultrasonicated samples due to bond breakage.

Furthermore, the XRD pattern of native and modified finger millet starch (WFM) showed characteristic A-type peaks. Annealing reduced the diffraction intensity and added a peak at 24.25°, suggesting crystalline reorientation. Hydrothermal modification introduced a V-type peak at 19.8°. The highest relative crystallinity (58.78%) was observed in AS due to better molecular realignment. In a study by Navyashree et al. (2022), the XRD analysis revealed differences in crystallinity among native (Nf), roasted (Rf), and germinated (Gf) white finger millet flours. All samples displayed A-type XRD patterns. Nf had the highest crystallinity (20.46%), followed by Gf (19.68%) and Rf (18.87%). Rf's decreased crystallinity was due to partial gelatinization during roasting, while Gf's reduction was caused by starch breakdown from amylase during germination (Olamiti et al., 2020; Sun et al., 2014).

Amarnath et al. (2023) presented scanning electron micrographs revealing that native WFM starch had polygonal, irregular, and oval shapes with surface indentations. Annealing (AS) caused pore development and granule fusion from partial gelatinization, while ultrasonication (US) created surface cracks and small holes due to high-pressure gradients. Both AS and US significantly altered starch morphology, consistent with changes in crystalline and amorphous regions. Similarly, another study by Arzoo et al. (2024) showed disoriented granules in the control millet, with increased disorientation in popped and parboiled samples. Parboiling resulted in merged granules while popping caused large gaps due to starch gelatinization, enhancing water absorption capacity. Soaked and germinated millet had rounder, polygonal structures with increased gaps, further loosening the structure and improving WAC and bulk density. Furthermore, Gowthamraj et al. (2021) studied the microstructural changes in finger millet flour under various treatments, including fermentation, lemon juice (T1), sodium chloride (T2), lemon juice followed by sodium chloride (T3), and fermentation followed by T3. The Nf showed irregular, compact clusters without surface pits. The control treatment resulted in hydrated granules, forming polygonal and spherical shapes. Lemon juice (FMFT1) caused surface corrosion and protein unfolding, leading to particle agglomeration. Sodium chloride (FMFT2), lemon juice with NaCl (FMFT3), and fermentation (FFMFT3) loosened starch particles, creating polygonal shapes with surface concavities. Fermentation also led to surface deterioration and pits due to enzymatic hydrolysis.

7.4.3 Protein and Starch Digestibility

Protein digestibility measures how much nitrogen is absorbed after proteins are divided into peptides and amino acids by enzymes in the small intestine. Gowthamraj et al. (2021) investigated the impact of fermentation on the protein profile and microstructural properties. Their findings revealed that protein digestibility improved with increased fermentation time. After 24 hours, in vitro protein digestibility (IVPD) peaked at 89%, significantly higher than other fermented samples, which ranged from 59% to 80%. This increase is attributed to proteolytic enzymes breaking down complex proteins into smaller, more digestible amino acids. In contrast, whole finger millet flour had a digestibility of around 53%, which decreased to 17.5%, 12.92%, and 13.40% after treatments with lemon juice (T1), NaCl solution (T2), and fermentation followed by treatment T3, respectively. These treatments likely removed the more digestible protein fractions. Chinma et al. (2022) examined the protein digestibility of germinated finger millet flour and Bambara groundnut flour blends-based bakery. Biscuits made with 100% germinated finger millet flour had a protein digestibility of 80.52%. As the proportion of germinated Bambara groundnut flour increased from 10% to 40%, protein digestibility rose to 89.20%. Slowly digestible starch (69.85–74.39%) and RS (7.21–14.90%) also increased with the addition of groundnut flour, whereas the levels of RDS, starch hydrolysis index, and glycemic index declined.

Onyango et al. (2020) examined the IVSD and IVPD of native finger millet flour (NFMF), hydrothermally treated finger millet flour (HTT-FMF), and composite flour bread with a 30% substitution of HTT-FMF. Their results showed a significant 9.71% decrease in digestible starch content in HTT-FMF, whereas RS rose from 11.3 g/100 g to 13.3 g/100 g due to the reassociation of starch polymers. Moreover, the protein digestibility decreased from 89.2% to 83.6% after hydrothermal treatment, and IVPD of composite bread (30% NFMF or HTT-FMF) ranged from 80.3% to 81.7%, slightly lower than control bread (82.5%) because of higher anti-nutritional factors available.

Puffing also affected its digestibility. Roopa and Premavalli (2008) observed that puffing significantly increased the RDS proportion in finger millet while simultaneously reducing SDS and RS levels. The starch digestibility of puffed millet, ranging from 29.6 to 35.8 mg/g, was observed to be four times greater than that of raw millet, which had a digestibility of only 7.2–9.9 mg/g. This significant increase in digestibility suggests that starch gelatinization has occurred during the puffing process. Devi and Rajendran (2021) reported that fermenting dosa batter with finger millet and horse gram improved IVSD from 86.9% to 90.3%, further rising to 92.3% after cooking due to increased enzymatic activity.

Furthermore, germination is found to increase IVPD significantly. Hejazi and Orsat (2016) found that germination significantly improved the IVPD of finger millet, with digestibility rising from 74% in ungerminated millet to 91% after 48 hours of germination at 30°C. Similarly, Syeunda et al. (2021) studied the effect of compositing precooked cowpea with malted finger millet on the IVPD of complementary porridge. Results showed an 11–18% increase in IVPD, with values rising from 73.28–76.17% in unmalted to 86.64–88.42% in malted KNE741 finger millet, attributed to partial protein hydrolysis.

7.5 Finger Millet-Based Food Products and Their Health Benefits

Finger millet is increasingly recognized for its significant nutritional value and health benefits, making it a vital component of various food products, especially in regions like India, where it is often referred to as "poor man's

TABLE 7.4

Finger Millet-Based Traditional and Commercial Products

S. No	Finger millet-based traditional food	Finger millet-based commercial food
1.	Salad	Puffed finger millet
2.	Idli	Multi millet chapati mix
3.	Porridge	Millet crisp, biscuits
4.	Khichidi	Millet noodle
5.	Chapati	Paysam mix
6.	Pancake	Muffin
7.	Pongal	Beverage

Source: Yousaf et al. (2021).

food" due to its affordability and high nutrient density (Dhanushkodi et al., 2023). This ancient grain is particularly rich in essential nutrients, including approximately 0.38% calcium, which is crucial for bone health; about 18% dietary fiber, beneficial for digestive health and glycemic control; and polyphenolic compounds ranging from 0.3% to 3%, known for their antioxidant properties (Devi et al., 2014). Finger millet-based traditional and commercial products are illustrated in Table 7.4.

7.5.1 Food Product

Finger millet-based products have garnered significant attention in the food industry for their versatile applications, offering numerous health benefits and the potential for developing value-added items that cater to a wide range of dietary needs. The flour derived from finger millet can be utilized to prepare nutrient-dense recipes suitable for both toddlers and adults, offering health advantages such as weight management, improved bone health, reduced blood cholesterol levels, and treatment of anemia (Jacob et al., 2024). In India, finger millet is commonly ground into flour and used to prepare traditional foods such as roti (unleavened flatbreads), kazhi (finger millet balls), and kanji (thin porridge). Additionally, finger millet undergoes various processing techniques to create popped, malted, extruded, and fermented products (Shobana et al., 2013).

Beyond its culinary applications, finger millet's bioactive compounds such as including tannins, steroids, polyphenols, alkaloids, terpenoids, cardiac glycosides, balsams, lignans, phytoestrogens, phytocyanin, gallic acid, ferulic acid, quercetin, vanillic acid, caffeic acid, sinapic acid, and proanthocyanidins can be extracted for medicinal purposes. These compounds hold potential for various therapeutic applications that contribute to overall health and well-being (Vinutha et al., 2024). Finger millet has been combined with other ingredients to create innovative biscuit formulations. For human nutrition purposes, it is typically used to make roti and other dishes. Several studies have suggested that incorporating finger millet flour with wheat flour in varying proportions can enhance the quality of baked goods such as bread and cookies (Kalsi et al., 2023; Krishnan et al., 2011). Kumari et al. (2023) found that incorporating encapsulated spray-dried ascorbic acid and iron gluconate powder into apple pomace and finger millet-based pasta can significantly enhance its nutritional value. The most suitable formulation (P1) included 10% finger millet, 4% apple pomace powder, 2% hydroxypropyl methylcellulose, and 84% durum semolina. Notably, the P1 pasta fortified with 0.5% encapsulated micronutrients retained an iron content of 23.74 mg/100 g and ascorbic acid content of 87.51 mg/100 g after cooking, aligning with the recommended dietary allowances and providing essential nutrients for adolescents.

7.5.2 Health Benefits

Numerous studies have highlighted the wide range of health benefits of consuming finger millet-based products. Its significantly elevated fiber quantity, along with the amino acid tryptophan, supports weight management by enhancing satiety, slowing digestion, and regulating blood sugar levels. This combination can lead to lower calorie intake, supporting effective and sustainable weight loss (Jacob et al., 2024). Additionally, it has calcium levels that are 5 to 30 times greater than those in other grains, making it essential for strengthening bones and teeth, which is especially advantageous for children and the elderly, as it supports their unique nutritional needs. For children, adequate nutrient intake is essential for growth and development. At the same time, older adults require specific nutrients to maintain bone density and overall health. By enhancing bone strength and mineral density, this nutrient-rich food helps reduce the likelihood of osteoporosis and fractures, which are significant concerns for aging populations. It contains a low glycemic index product, which, paired with its fiber proportion, helps maintain healthy blood pressure and supports vascular health by increasing blood vessel elasticity and relaxation, promoting better circulation and cardiovascular health (Sahaya Rani et al., 2021).

Finger millet is abundant in iron and calcium, which play a crucial role in enhancing hemoglobin levels and helping to prevent iron deficiency anemia, especially in children and adolescents. Moreover, its nutritional benefits assist as a prebiotic, which is crucial in helping a nourishing gut microbiome. Fostering the growth and movement of beneficial probiotics helps maintain a balanced intestinal environment. This support for gut health is essential, as a thriving microbiome can enhance digestion, improve nutrient absorption, and strengthen the immune system. This dual action not only supports overall health but also contributes to enhanced digestive wellness (Dhanushkodi et al., 2023). It provides various therapeutic benefits, primarily attributed to its ability to combat oxidative stress, reduce inflammation, and inhibit microbial growth. These characteristics can significantly help mitigate health risks associated with tumors, cardiovascular diseases, and metabolic syndrome. Furthermore, the grain's hypoglycemic and hypolipidemic effects make it an excellent food choice for individuals looking to manage diabetes and high cholesterol levels effectively. By incorporating finger millet into their diet, individuals can not only enjoy its nutritional advantages but also promote overall health and well-being (Jacob et al., 2024).

7.6 Conclusion

Finger millet offers substantial potential for enhancing food security and supporting sustainable development, particularly in regions where it has been traditionally cultivated. Its abundance in dietary fiber, RS, and essential minerals offers a highly nutritious food choice that promotes better health outcomes. However, finger millet remains underutilized despite its abundant starch content, largely due to limited awareness. However, with its gluten-free nature and unique nutritional properties, is gaining increasing attention, particularly among individuals with celiac disease. Its abundant proportion of dietary fiber, RS, and essential minerals, alongside being gluten-free, makes it a versatile and health-enhancing ingredient. While its potential remains underutilized due to limited awareness, various processing methods are being explored to unlock its value. These methods significantly enhance finger millet's functional, antioxidant, and structural qualities, positioning it as a competitive ingredient in the food industry. By modifying finger millet through techniques that boost its nutritional and functional properties, the grain becomes a prime candidate for innovative, health-driven food products. This opens up further additional opportunities for its use beyond the food industry, extending into cosmetics, pharmaceuticals, and biodegradable packaging, thereby broadening its impact and enhancing its contribution to sustainable development.

References

Abah, C. R., Ishiwu, C. N., Obiegbuna, J. E., & Oladejo, A. A. (2020). Nutritional composition, functional properties and food applications of millet grains. *Asian Food Science Journal, 14*(2), 9–19.

Abioye, V. F., Babarinde, G. O., Ogunlakin, G. O., Adejuyitan, J. A., Olatunde, S. J., & Abioye, A. O. (2022). Varietal and processing influence on nutritional and phytochemical properties of finger millet: A review. *Heliyon, 8*(12), e12310. https://doi.org/10.1016/j.heliyon.2022.e12310

Aisoni, J. E., Yusha'U, M., & Orole, O. O. (2018). Processing effects on physicochemical and proximate composition of finger millet (*Eleusine coracana*). *Greener Journal of Biological Sciences*, *8*(2), 14–20. https://doi.org/10.15580/GJBS.2018.2.032018048

Amarnath, M. S., Muhammed, A., Antony, A. K., Malini, B., & Sunil, C. K. (2023). White finger millet starch: Physical modification (annealing and ultrasound), and its impact on physicochemical, functional, thermal and structural properties. *Food and Humanity*, *1*, 599–606. https://doi.org/10.1016/j.foohum.2023.07.010

Anal, A. K., Singh, R., Rice, D., Pongtong, K., Hazarika, U., Trivedi, D., & Karki, S. (2024). Millets as supergrains: A holistic approach for sustainable and healthy food product development. *Sustainable Food Technology*. https://doi.org/10.1039/d4fb00047a

Ananthu, R., Buvaneswaran, M., Meena, L., & C K, S. (2023). Microwave treatment effect on functional and pasting properties, and storage stability of white finger millet. *Journal of Food Process Engineering*, *46*. https://doi.org/10.1111/jfpe.14341

Annison, G., & Topping, D. L. (1994). Nutritional role of resistnt starch : Chemical structure vs physiological function primay structure of starch components in plants, starches are packaged in granules. Following disruption and disso. *Annual Review of Nutrition*, *14*(1), 297–320.

Antony, U., Sripriya, G., & Chandra, T. S. (1996). Effect of fermentation on the primary nutrients in finger millet (*Eleusine coracana*). *Journal of Agricultural and Food Chemistry*, *44*(9), 2616–2618. https://doi.org/10.1021/jf950787q

Antony Ceasar, S., Maharajan, T., Ajeesh Krishna, T. P., Ramakrishnan, M., Victor Roch, G., Satish, L., & Ignacimuthu, S. (2018). Finger millet [*Eleusine coracana* (L.) Gaertn.] Improvement: Current status and future interventions of whole genome sequence. *Frontiers in Plant Science*, *9*, 1054.

Arzoo, Neeru, Shukla, A. K., Panwar, S., & Kumar, A. (2024). Effect of processing techniques on functional, antioxidants, and structural properties of finger millet (*Eleusine coracana*) flour. *Food and Humanity*, *2*, 100305. https://doi.org/10.1016/j.foohum.2024.100305

Azeez, S. O., Chinma, C. E., Bassey, S. O., Eze, U. R., Makinde, A. F., Sakariyah, A. A., Okubanjo, S. S., Danbaba, N., & Adebo, O. A. (2022). Impact of germination alone or in combination with solid-state fermentation on the physicochemical, antioxidant, in vitro digestibility, functional and thermal properties of brown finger millet flours. *LWT*, *154*. https://doi.org/10.1016/j.lwt.2021.112734

Balakumaran, M., Gokul Nath, K., Giridharan, B., Dhinesh, K., Dharunbalaji, A. K., Malini, B., & Sunil, C. K. (2023). White finger millet starch: Hydrothermal and microwave modification and its characterization. *International Journal of Biological Macromolecules*, *242*. https://doi.org/10.1016/j.ijbiomac.2023.124619

Balasubramaniam, V. G., Ayyappan, P., Sathvika, S., & Antony, U. (2019). Effect of enzyme pretreatment in the ultrasound-assisted extraction of finger millet polyphenols. *Journal of Food Science and Technology*, *56*(3), 1583–1594. https://doi.org/10.1007/s13197-019-03672-2

Bhasin, A., Maibam, B. D., & Sharma, M. (2024). Review on Finger millet (*Eleusine coracana* L.): Nutritional composition, modification, its effect on physicochemical, structural, functional properties, and its applications. *Journal of Food Composition and Analysis*, 106623.

Chandra, D., Chandra, S., Pallavi, & Sharma, A. K. (2016). Review of Finger millet (*Eleusine coracana* (L.) Gaertn): A power house of health benefiting nutrients. *Food Science and Human Wellness, 5*(3), 149–155. https://doi.org/10.1016/j.fshw.2016.05.004

Chandrasekara, A., Naczk, M., & Shahidi, F. (2012). Effect of processing on the antioxidant activity of millet grains. *Food Chemistry, 133*(1), 1–9. https://doi.org/10.1016/j.foodchem.2011.09.043

Chandrasekara, A., Senanayake, I., Kumari, D., & Shahidi, F. (2022). Effect of processing on the antioxidant activities of porridges and pittu prepared from finger millets (*Eleusine coracana*). *Food Production, Processing and Nutrition, 4*(1). https://doi.org/10.1186/s43014-022-00097-x

Chauhan, E. S., & Sarita. (2018). Effects of processing (germination and popping) on the nutritional and anti-nutritional properties of finger millet (*Eleusine coracana*). *Current Research in Nutrition and Food Science, 6*(2), 566–572. https://doi.org/10.12944/CRNFSJ.6.2.30

Chinma, C. E., Ibrahim, P. A., Adedeji, O. E., Ezeocha, V. C., Ohuoba, E. U., Kolo, S. I., Abdulrahman, R., Ogochukwu Anumba, N. L., Adebo, J. A., & Adebo, O. A. (2022). Physicochemical properties, in vitro digestibility, antioxidant activity and consumer acceptability of biscuits prepared from germinated finger millet and Bambara groundnut flour blends. *Heliyon, 8*(10). https://doi.org/10.1016/j.heliyon.2022.e10849

Devi, P. B., & Rajendran, S. (2021). Impact of starter culture on nutraceutical and functional properties of underutilized millet-legume co-fermented Indian traditional product. *LWT, 149*. https://doi.org/10.1016/j.lwt.2021.111818

Devi, P. B., Vijayabharathi, R., Sathyabama, S., Malleshi, N. G., & Priyadarisini, V. B. (2014). Health benefits of finger millet (*Eleusine coracana* L.) polyphenols and dietary fiber: A review. *Journal of Food Science and Technology, 51*(6), 1021–1040. https://doi.org/10.1007/s13197-011-0584-9

Dhanushkodi, V., Hemavathy, A. T., Shenbagavalli, S., Sangeetha, S., Anitha, R., & UmaMaheshwari, T. (2023). A review on nutritional properties and health benefits of finger millet. *International Journal of Plant & Soil Science, 35*(18), 753–761. https://doi.org/10.9734/ijpss/2023/v35i183342

Dharmaraj, U., & Malleshi, N. G. (2011). Changes in carbohydrates, proteins and lipids of finger millet after hydrothermal processing. *LWT, 44*(7), 1636–1642. https://doi.org/10.1016/j.lwt.2010.08.014

Dharmaraj, U., Meera, M. S., Reddy, S. Y., & Malleshi, N. G. (2015). Influence of hydrothermal processing on functional properties and grain morphology of finger millet. *Journal of Food Science and Technology, 52*(3), 1361–1371. https://doi.org/10.1007/s13197-013-1159-8

Dhliwayo, T., Chopera, P., Matsungo, T. M., Chidewe, C., Mukanganyama, S., Nyakudya, E., Mtambanengwe, F., Mapfumo, P., & Nyanga, L. K. (2023). Effect of germination and roasting on the proximate, mineral and anti-nutritional factors in finger millet (*Eleucine coracana*), cowpeas (*Vigna unguiculata*) and orange maize (*Zea mays*). *African Journal of Food, Agriculture, Nutrition and Development, 23*(8), 24346–24362. https://doi.org/10.18697/ajfand.123.22960

Dias-Martins, A. M., Cappato, L. P., da Costa Mattos, M., Rodrigues, F. N., Pacheco, S., & Carvalho, C. W. P. (2019). Impacts of ohmic heating on decorticated and whole pearl millet grains compared to open-pan cooking. *Journal of Cereal Science, 85*, 120–129. https://doi.org/10.1016/j.jcs.2018.11.007

Dubey, A., & Tripathy, P. P. (2024). Ultrasound-mediated hydration of finger millet: Effects on antinutrients, techno-functional and bioactive properties, with evaluation of ANN-PSO and RSM optimization methods. *Food Chemistry, 435.* https://doi.org/10.1016/j.foodchem.2023.137516

Gabaza, M., Shumoy, H., Muchuweti, M., Vandamme, P., & Raes, K. (2016). Effect of fermentation and cooking on soluble and bound phenolic profiles of finger millet sour porridge. *Journal of Agricultural and Food Chemistry, 64*(40), 7615–7621. https://doi.org/10.1021/acs.jafc.6b03090

George, N., Mildred, N., Wanzala, E., Munga, J., Oduori, C., Kinyuru, J., & Hudson, N. (2023). Nutritional composition and anti-nutrient levels in raw and processed varieties of finger millet promoted for nutritional security. *Food and Nutrition Sciences, 14*(12), 1183–1205. https://doi.org/10.4236/fns.2023.1412074

Gowda, N. A. N., Siliveru, K., Prasad, P. V. V., Bhatt, Y., Netravati, B. P., & Gurikar, C. (2022). Modern processing of Indian millets: A perspective on changes in nutritional properties. *Foods, 11*(4), 499.

Gowthamraj, G., Raasmika, M., & Narayanasamy, S. (2021). Efficacy of fermentation parameters on protein quality and microstructural properties of processed finger millet flour. *Journal of Food Science and Technology, 58*(8), 3223–3234. https://doi.org/10.1007/s13197-020-04826-3

Hejazi, S. N., & Orsat, V. (2016). Malting process optimization for protein digestibility enhancement in finger millet grain. *Journal of Food Science and Technology, 53*(4), 1929–1938. https://doi.org/10.1007/s13197-016-2188-x

Hithamani, G., & Srinivasan, K. (2014). Effect of domestic processing on the polyphenol content and bioaccessibility in finger millet (*Eleusine coracana*) and pearl millet (*Pennisetum glaucum*). *Food Chemistry, 164,* 55–62. https://doi.org/10.1016/j.foodchem.2014.04.107

Jacob, J., Krishnan, V., Antony, C., Bhavyasri, M., Aruna, C., Mishra, K., Nepolean, T., Satyavathi, C. T., & Visarada, K. B. R. S. (2024). The nutrition and therapeutic potential of millets: An updated narrative review. *Frontiers in Nutrition, 11,* 1–31. https://doi.org/10.3389/fnut.2024.1346869

Jan, S., Kumar, K., Yadav, A. N., Ahmed, N., Thakur, P., Chauhan, D., Rizvi, Q. U. E. H., & Dhaliwal, H. S. (2022). Effect of diverse fermentation treatments on nutritional composition, bioactive components, and anti-nutritional factors of finger millet (*Eleusine coracana* L.). *Journal of Applied Biology and Biotechnology, 10,* 46–52. https://doi.org/10.7324/JABB.2022.10s107

Jayawardana, S. A. S., Samarasekera, J. K. R. R., Hettiarachchi, G. H. C. M., Gooneratne, J., Mazumdar, S. D., & Banerjee, R. (2019). Dietary fibers, starch fractions and nutritional composition of finger millet varieties cultivated in Sri Lanka. *Journal of Food Composition and Analysis, 82*(363), 103249. https://doi.org/10.1016/j.jfca.2019.103249

Jayawardana, S. A. S., Samarasekera, J. K. R. R., Hettiarachchi, G. H. C. M., & Gooneratne, M. J. (2022). Antidiabetic properties of finger millet (*Eleusine coracana* (L.) Gaertn.) varieties cultivated in Sri Lanka. *Journal of Herbal Medicine, 32,* 100534.

Joshi, T. J., Singh, S. M., & Rao, P. S. (2023). Novel thermal and non-thermal millet processing technologies: Advances and research trends. *European Food Research and Technology, 249*(5), 1149–1160. https://doi.org/10.1007/s00217-023-04227-8

Kalsi, R., Bhasin, J., Goksen, G., & Kashyap, P. (2023). Exploration of nutritional, pharmacological, and the processing trends for valorization of finger millet (*Eleusine coracana*): A review. *Food Science and Nutrition, 11*(11), 6802–6819. https://doi.org/10.1002/fsn3.3659

Kaur, B., Singh, A., Suri, S., Usman, M., & Dutta, D. (2023). Minor millets: A review on nutritional composition, starch extraction/modification, product formulation, and health benefits. *Journal of the Science of Food and Agriculture, 103*(10), 4742–4754.

Kaur, H., & Gill, B. S. (2021). Changes in physicochemical, nutritional characteristics and ATR–FTIR molecular interactions of cereal grains during germination. *Journal of Food Science and Technology, 58*(6), 2313–2324. https://doi.org/10.1007/s13197-020-04742-6

Krishnan, R., Dharmaraj, U., & Malleshi, N. G. (2012). Influence of decortication, popping and malting on bioaccessibility of calcium, iron and zinc in finger millet. *LWT, 48*(2), 169–174. https://doi.org/10.1016/j.lwt.2012.03.003

Krishnan, R., Dharmaraj, U., Sai Manohar, R., & Malleshi, N. G. (2011). Quality characteristics of biscuits prepared from finger millet seed coat based composite flour. *Food Chemistry, 129*(2), 499–506. https://doi.org/10.1016/j.foodchem.2011.04.107

Kumar, A., Kaur, A., Gupta, K., Gat, Y., & Kumar, V. (2021). Assessment of germination time of finger millet for value addition in functional foods. *Current Science, 120*(2).

Kumar, S. R., Sadiq, M. B., & Anal, A. K. (2020). Comparative study of physicochemical and functional properties of pan and microwave cooked underutilized millets (proso and little). *LWT, 128*(September 2019), 109465. https://doi.org/10.1016/j.lwt.2020.109465

Kumar, S. R., Tangsrianugul, N., & Suphantharika, M. (2023). A review on isolation, characterization, modification, and applications of proso millet starch. *Foods, 12*(12), 1–16. https://doi.org/10.3390/foods12122413

Kumar, Y., Sharanagat, V. S., Singh, L., & Mani, S. (2020). Effect of germination and roasting on the proximate composition, total phenolics, and functional properties of black chickpea (Cicer arietinum). *Legume Science, 2*(1), e20.

Kumari, M., Arya, P., Khatkar, S. K., & Kumar, P. (2023). Development and characterization of apple pomace and finger millet-based pasta enriched with encapsulated micronutrient. *Food Chemistry Advances, 3*(June), 100551. https://doi.org/10.1016/j.focha.2023.100551

Liu, P., Cheng, J., Li, M., Li, J., Zhu, H., Wang, S., & Yu, G. (2020). Effects of extrusion conditions on physicochemical properties of finger millet (*Gramineae setaria*). *Journal of Agricultural Science, 12*(9), 149. https://doi.org/10.5539/jas.v12n9p149

Makokha, A. O., Oniang'o, R. K., Njoroge, S. M., & Kamar, O. K. (2002). Effect of traditional fermentation and malting on phytic acid and mineral availability from sorghum (*Sorghum bicolor*) and finger millet (*Eleusine coracana*) grain varieties grown in Kenya. *Food and Nutrition Bulletin, 23*(3_suppl1), 241–245.

Mwanyigha, W. M., Muriithi, A. N., & Konyole, S. O. (2023). Acceptability and nutritional quality of finger millet (Eleusine coracana) cookies enriched with cricket (Acheta domesticus). Journal of Food Stability, 6(2), 84–95. DOI: 10.36400/J.Food.Stab.6.2.2023-010

Mutshinyani, M., Mashau, M. E., & Jideani, A. I. O. (2020). Bioactive compounds, antioxidant activity and consumer acceptability of porridges of finger millet (*Eleusine coracana*) flours: Effects of spontaneous fermentation. *International Journal of Food Properties, 23*(1), 1692–1710. https://doi.org/10.1080/10942912.2020.1825485

Nanje Gowda, N. A., Siliveru, K., Vara Prasad, P. V., Bhatt, Y., Netravati, B. P., & Gurikar, C. (2022). Modern processing of Indian millets: A perspective on changes in nutritional properties. *Foods, 11*(4). https://doi.org/10.3390/foods11040499

Navyashree, N., Singh Sengar, A., Sunil, C. K., & Venkatachalapathy, N. (2022). White finger millet (KMR-340): A comparative study to determine the effect of processing and their characterisation. *Food Chemistry, 374,* 131665. https://doi.org/10.1016/j.foodchem.2021.131665

Nishani, S., Rudra, S., & Verghese, E. (2017). Effect of extrusion processing variables on finger millet flours with respect to their functional properties using response surface methodology. *Current Journal of Applied Science and Technology, 23*(2), 1–9. https://doi.org/10.9734/cjast/2017/34495

Olamiti, G., Takalani, T. K., Beswa, D., & Jideani, A. I. O. (2020). Effect of malting and fermentation on colour, thermal properties, functional groups and crystallinity level of flours from pearl millet (*Pennisetum glaucum*) and sorghum (*Sorghum bicolor*). *Heliyon, 6*(12). https://doi.org/10.1016/j.heliyon.2020.e05467

Onweluzo, J. C., & Nwabugwu, C. C. (2009). Fermentation of millet and pigeon pea seeds for flour production: Effects of composition and selected functional properties. *Pakistan Journal of Nutrition, 8*(6), 737–744.

Onyango, C., Luvitaa, S. K., Unbehend, G., & Haase, N. (2020). Physico-chemical properties of flour, dough and bread from wheat and hydrothermally-treated finger millet. *Journal of Cereal Science, 93.* https://doi.org/10.1016/j.jcs.2020.102954

Panghal, A., Khatkar, B. S., Yadav, D. N., & Chhikara, N. (2019). Effect of finger millet on nutritional, rheological, and pasting profile of whole wheat flat bread (chapatti). *Cereal Chemistry, 96*(1), 86–94. https://doi.org/10.1002/cche.10111

Patel, K. K., Naik, R. K., Patel, G., & Pandey, S. (2022). Studies on some physical and engineering properties of finger millet for designing thresher. *Biological Forum, 14*(4), 387–393.

Patil, P., Singh, S. P., & Patel, P. (2023). Functional properties and health benefits of finger millet (*Eleusine coracana* L.): A review. *The Journal of Phytopharmacology, 12*(3), 196–202. https://doi.org/10.31254/phyto.2023.12308

Patil, S. S., Rudra, S. G., Varghese, E., & Kaur, C. (2016). Effect of extruded finger millet (*Eleusine coracana* L.) on textural properties and sensory acceptability of composite bread. *Food Bioscience, 14,* 62–69. https://doi.org/10.1016/j.fbio.2016.04.001

Patil, S. S., Varghese, E., Rudra, S. G., & Kaur, C. (2016). Effect of extrusion processing on phenolics, flavonoids and antioxidant activity of millets. *International Journal of Food and Fermentation Technology, 6*(1), 177. https://doi.org/10.5958/2277-9396.2016.00040.4

Pradeep, P. M., & Sreerama, Y. N. (2015). Impact of processing on the phenolic profiles of small millets: Evaluation of their antioxidant and enzyme inhibitory properties associated with hyperglycemia. *Food Chemistry, 169,* 455–463.

Ramashia, S. E., Anyasi, T. A., Gwata, E. T., Meddows-Taylor, S., & Jideani, A. I. O. (2019). Processing, nutritional composition and health benefits of finger millet sub-Saharan Africa. *Food Science and Technology (Brazil), 39*(2), 253–266. https://doi.org/10.1590/fst.25017

Rasri, W., Kumar, S. R., & Khampratueng, P. (2024). Valorization of ragi husk and its utilization in the food packaging industry. *Agro-Wastes for Packaging Applications, 83.*

Rathore, T., Singh, R., Kamble, D. B., Upadhyay, A., & Thangalakshmi, S (2019). Review on finger millet: Processing and value addition. *The Pharma Innovation Journal, 8*(4), 283–291.

Roopa, S., & Premavalli, K. S. (2008). Effect of processing on starch fractions in different varieties of finger millet. *Food Chemistry, 106*(3), 875–882. https://doi.org/10.1016/j.foodchem.2006.08.035

Sahaya Rani, G., Swaminathan, A., & Vijayaraghavan, R. (2021). Effectiveness of physical activity and finger millet-based food supplement on biochemical parameters and bone mineral density among premenopausal women. *Evidence-Based Complementary and Alternative Medicine, 2021.* https://doi.org/10.1155/2021/4757991

Samtiya, M., Soni, K., Chawla, S., Poonia, A., Sehgal, S., & Dhewa, T. (2021). Key antinutrients of millet and their reduction strategies: An overview. *Acta Scientific Nutritional Health, 5*(12).

Sarita, E. S., & Singh, E. (2016). Potential of millets: Nutrients composition and health benefits. *Journal of Scientific and Innovative Research, 5*(2), 46–50.

Sharanagat, V. S., Suhag, R., Anand, P., Deswal, G., Kumar, R., Chaudhary, A., Singh, L., Singh Kushwah, O., Mani, S., Kumar, Y., & Nema, P. K. (2019). Physico-functional, thermo-pasting and antioxidant properties of microwave roasted sorghum [*Sorghum bicolor* (L.) Moench]. *Journal of Cereal Science, 85,* 111–119. https://doi.org/10.1016/j.jcs.2018.11.013

Sharma, S., & Garg, A. P. (2023). Effect of germination on the physicochemical and antinutritional properties of finger millet (*Eleusine coracana*), pearl millet (*Pennisetum glaucum*), and sorghum (*Sorghum bicolor*). *The Pharma Innovation Journal, 12*(3), 4763–4772. www.thepharmajournal.com

Shobana, S., Krishnaswamy, K., Sudha, V., Malleshi, N. G., Anjana, R. M., Palaniappan, L., & Mohan, V. (2013). Finger millet (ragi, *Eleusine coracana* L.). A review of its nutritional properties, processing, and plausible health benefits. In *Advances in Food and Nutrition Research* (1st ed., Vol. 69). Elsevier. https://doi.org/10.1016/B978-0-12-410540-9.00001-6

Shobana, S., Sreerama, Y. N., & Malleshi, N. G. (2009). Composition and enzyme inhibitory properties of finger millet (*Eleusine coracana* L.) seed coat phenolics: Mode of inhibition of α-glucosidase and pancreatic amylase. *Food Chemistry, 115*(4), 1268–1273. https://doi.org/10.1016/j.foodchem.2009.01.042

Shukla, A. K., Panwar, S., & Kumar, A. (2024). Effect of processing techniques on functional, antioxidants, and structural properties of finger millet (*Eleusine coracana*) flour. *Food and Humanity, 2,* 100305.

Singh, A., Gupta, S., Kaur, R., & Gupta, H. R. (2017). Process optimization for antinutrient minimization of millets. *Asian Journal of Dairy and Food Research.* https://doi.org/10.18805/ajdfr.dr-1215

Singh, M., Manickavasagan, A., Shobana, S., & Mohan, V. (2021). Glycemic index of pulses and pulse-based products: A review. *Critical Reviews in Food Science and Nutrition, 61*(9), 1567–1588.

Singh, N., David, J., Thompkinson, D., Sagar Seelam, B., Rajput, H., & Morya, S. (2018). Effect of roasting on functional and phytochemical constituents of finger millet (*Eleusine coracana* L.). *The Pharma Innovation Journal, 7*(4), 414–418. www.thepharmajournal.com

Singh, S. M., Jayasree Joshi, T., & Rao, P. S. (2024). Technological advancements in millet dehusking and polishing process; an insight into pretreatment methods, machineries and impact on nutritional quality. In *Grain and Oil Science and Technology.* KeAi Communications Co. https://doi.org/10.1016/j.gaost.2024.05.007

Sridhar, G. L. (1994). Contents of total lipids and lipid classes and composition of fatty acids in small millets: foxtail (*Setaria italica*), proso (*Panicum miliaceum*), and finger (*Eleusine coracana*). *Cereal Chemistry, 71*, 355–359.

Sruthi, N. U., & Rao, P. S. (2021). Effect of processing on storage stability of millet flour: A review. *Trends in Food Science and Technology, 112*, 58–74. https://doi.org/10.1016/j.tifs.2021.03.043

Sun, Q., Han, Z., Wang, L., & Xiong, L. (2014). Physicochemical differences between sorghum starch and sorghum flour modified by heat-moisture treatment. *Food Chemistry, 145*, 756–764. https://doi.org/10.1016/j.foodchem.2013.08.129

Syeunda, C. O., Anyango, J. O., Faraj, A. K., & Kimurto, P. K. (2021). In vitro protein digestibility of finger millet complementary porridge as affected by compositing precooked cowpea with improved malted finger millet. *Journal of Food Science and Technology, 58*(2), 571–580. https://doi.org/10.1007/s13197-020-04569-1

Tagade, A., & Sawarkar, A. N. (2023). Valorization of millet agro-residues for bioenergy production through pyrolysis: Recent inroads, technological bottlenecks, possible remedies, and future directions. *Bioresource Technology, 384*, 129335. https://doi.org/10.1016/J.BIORTECH.2023.129335

Tharanathan, R. N., & Mahadevamma, S. (2003). Grain legumes – A boon to human nutrition. *Trends in Food Science and Technology, 14*(12), 507–518. https://doi.org/10.1016/j.tifs.2003.07.002

Tonapi, V. A., Ganapathy, K. N., Hariprasanna, K., Bhat, B. V., Amasiddha, B., Avinash, S., & Deepika, C. (2022). Small Millets breeding. In *Fundamentals of Field Crop Breeding*. https://doi.org/10.1007/978-981-16-9257-4_8

Tripathi, M. K., Kumar, A., Mohapatra, D., Jadam, R. S., Selvan, S. S., & Nickhil, C. (2022). Economic, nutritional, and health importance of finger millet (pp.13–33). https://doi.org/10.1007/978-3-031-00868-9_2

Vinutha B. N., Soumya C. Meti, and S. R. (2024). Finger millet: Biology, functional potential and sustainable utilization major millets. In *Millets: The Multi-Cereal Paradigm for Food Sustainability, World Sustainability Series.*

Wankhede, D. B., Rathi, S. S., Gunjal, B. B., Patil, H. B., Walde, S. G., Rodge, A. B., & Sawate, A. R. (1990). Studies on isolation and characterization of starch from pearl millet (*Pennisetum americanum* (L.) Leeke) grains. *Carbohydrate Polymers, 13*(1), 17–28.

Yadav, S., Mishra, S., & Pradhan, R. C. (2021). Ultrasound-assisted hydration of finger millet (*Eleusine Coracana*) and its effects on starch isolates and antinutrients. *Ultrasonics Sonochemistry, 73*. https://doi.org/10.1016/j.ultsonch.2021.105542

Yenasew, A., & Urga, K. (2022). Effect of germination period on physiochemical properties of elite finger millet varieties. *Cogent Food and Agriculture, 8*(1). https://doi.org/10.1080/23311932.2022.2093045

Yousaf, L., Hou, D., Liaqat, H., & Shen, Q. (2021). Millet: A review of its nutritional and functional changes during processing. *Food Research International, 142*(17), 110197. https://doi.org/10.1016/j.foodres.2021.110197

8

Nutraceutical and Therapeutic Aspects of Finger Millet

Shiv and Rinku

8.1 Introduction

Millets are one of the oldest crops grown in India and is constantly attracting the current population to fulfil their requirement of a nutrient-wealthy food regimen. Millets are rich in dietary fibre, proteins, micronutrients, and bioactive compounds and also exhibit a lower glycaemic index. Owing to their exceptional nutritional composition these ancient staples are also termed as nutraceuticals.

Finger millet (*Eleusine corocana*) is a crucial millet crop with excessive organic activity, nutritive, and healing properties. Finger millet is gaining the interest of the modern generation as the health benefits of finger millet come out with new research and experiments. Finger millet has high nutrient density as it contains carbohydrates, proteins, vitamins and minerals. It contains about 65–75% carbohydrates, 5–8% proteins, 1–2% fat, 15–20% dietary fibre and 2.5–3.5% minerals with an energy content of 300–350 kcal (Kaur et al., 2024). It is also rich in polyphenol compounds, flavonoids, phytochemicals and tannins which enhances its uses in industrial applications (Nayak & Satish, 2024). It has been proved that it shows higher digestibility with least allergenicity. It is anti-diabetic as it reduces the glycaemic index due to slow digestibility (Rathore & Sharma, 2012). Improvement in the diet is required to avoid diseases caused by metabolic disorders such as obesity and type II diabetes (Ashik Somarajan & Morya, 2022). Ragi is an excellent source of calcium (0.34%), which plays a pivotal role in strengthening of bones and reducing the chances of osteoporosis (Dube, Varade, & Choudhari, 2024; Warhade, Atram, & Sawarkar, 2023). Attributing to its low glycaemic index, it aids in controlling diabetes by activating insulin (Bhujle, Nayak, Gowda, Pandiselvam, & Sunil, 2024); also reduces skin ageing (Meti & Rawat, 2024); and makes skin tissues less prone to sagging and wrinkled appearance (Shrestha et al., 2024). It is also one of the rarest natural sources of vitamin D, a superb source of iron, which in turn assists in treating anaemia and can

DOI: 10.1201/9781003535119-8

also help in the management of cognitive diseases, anxiety, depression and insomnia (Choudhury, Chaudhary, & Research, 2024; Premachandran, Wilson Alphonse, Soundharapandiyan, & Research, 2023; Tanwar & Kalpana, 2024). Additionally, it decreases the intensity of asthmatic attacks and impart a positive impact on migraines (Bisht, Bisht, Gudadhe, Raut, & Dobhal, 2023; Kavitha, Nihar, & Kader, 2024; Patil, Singh, & Patel, 2023). Finger millet is known as a remedy for leprosy and liver diseases and can also help in the treatment of measles, pneumonia and smallpox (Chand, Rajnish, Chandra, Dwivedi, & Singh, 2024; Parveen et al., 2025).

It may be grown with special cultivars relying on using millet. The white cultivars are grown to be used in the baking industry, mild brown cultivars may be utilized in porridge, and brown cultivars may be used for brewing in beer industry (Vellaiyan et al., 2023). Biological activities of finger millet include anti-diabetic, anticarcinogenic, antioxidant, anti-ageing, and cardio-protective impact (Ashik Somarajan & Morya, 2022).

Along with these biological activities, it has found tremendous advantages in industrial applications such as the development of functional food with the addition of millet, bio-energy production and nutraceutical applications. The ease of storage of finger millets and extended shelf life makes it a reliable food source during periods of drought and food scarcity, further underlining its importance in traditional diets and medicinal practices. Finger millet can be added to different food materials to prepare infant foods, beverages like ragi malt, fermented foods, bakery foods, etc., which enhances fibre content, increases calcium and protein content and reduces antinutritional content. It is used to prepare noodles which decrease glycaemic index and prove a good food replacer for diabetic patients. If the seed coat of finger millet is to be separated, then it can be used as an adjunct in the preparation of special foods (Devi, Vijayabharathi, Sathyabama, Malleshi, & Priyadarisini, 2014). Along with these benefits, finger millet can be added to various dairy and food products to increase their nutritional and therapeutic properties (Ashik Somarajan & Morya, 2022). This chapter will cover topics such as the health benefits and therapeutic effects of finger millet with recent case studies.

8.2 Health Benefits of Finger Millet

In developing countries, malnutrition is a major problem in children and lactating women because of a lack of protein-rich diets and supplements. According to WHO, malnutrition is caused by to deficiency of vitamin A, iron, iodine and protein. Millets are rich sources of nutrients like vitamins, minerals, proteins and dietary fibre which proved as a remedy for the malnutrition problems in growing populations (Vellaiyan et al., 2023). Finger millet contains a higher amount of dietary fibre than other millet as it has

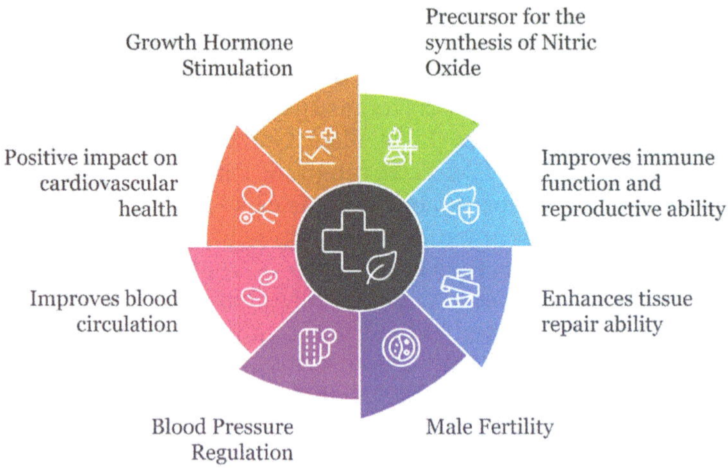

FIGURE 8.1
Health benefits of finger millet.

five layered seed coat (Kaur et al., 2024). Whereas, carbohydrates give a low glycaemic index as it contain slowly digestible starch, dietary fibre, and resistant starch. Finger millet contains about 1.92% fat, which comprises 74.4% unsaturated fatty acid and 25.6% saturated fatty acids (Bisht, Bisht, Gudadhe, Raut, & Dobhal, 2023). According to (Kumar et al., 2016) studies, it was found that plant-based food mainly whole grain cereals helps to prevent many severe disorders including diabetes, tumour, cardiovascular diseases (CVDs), obesity problems and unhealthy gut microbiota, etc. To prevent these diseases, various studies have shown the importance of finger millet to tackle these problems as it contains various phenolic compounds, minerals, vitamins proteins, etc. These components can be separated and used as nutraceuticals and functional food formulations. Millets contain several essential amino acids that provide various benefits as depicted in Fig. 8.1 (Vellaiyan et al., 2023).

8.2.1 Diabetes Management

In the current scenario, with increased technological access, humans tend to follow a more sedentary and stressful life. The world is becoming a hub for diseases like diabetes mellitus, obesity, CVDs, hypertension and stroke in the much lower age groups, even in 30s and 40s (Mansoria, Singh, & Sciences, 2023). Diabetes is characterized by a complex pathogenesis. It can be type-1 (autoimmune disorder, inadequate production of insulin) and type-2 (non-insulin-dependent, results either from insulin resistance or beta cell dysfunction). In addition to the above, exhaustion of insulin transporters in the kidney and intestine may also lead to diabetes.

Diabetes is a lifestyle-related disorder that can be managed through dietary intervention along with medications (Owheruo et al., 2024). The amalgamation of millets with locally available staple foods and vegetables in the diet bestows significant benefits in the management of diabetes. A study focusing on the awareness of millet utilization and diabetes management was conducted in Visakhapatnam, Andhra Pradesh. A total of 112 random respondents were selected from four zones (28 participants in each zone). The team used both primary (personal interactions) and secondary data (collected from journals, books etc.) for analysis. The results revealed that a good number of participants were managing their diabetes by inculcating millet in their diet two times a day.

Methanolic extracts of finger millet inhibit collagen glycation which is beneficial for the human body. Diabetes can be controlled and regulated by consuming calcium and magnesium in a daily diet (Kumar et al., 2016). Polyphenols present in finger millet show a synergistic effect with dietary fibre which helps in the inhibition of amylase, contributing to reducing the risk of type II diabetes. Vanillic acid, present in finger millet, is a dihydroxybenzoic acid derivative and shows a nephroprotective effect against diabetes. Synthetic inhibitors of alpha-glucosidase and pancreatic amylase play a very important role in the treatment of diabetes by controlling post-meal hyperglycaemia. These inhibitory compounds act by reducing the catabolism of complex carbohydrates leading to sustained release of glucose into the bloodstream. It was found that polyphenols present in finger millet are responsible for reducing the problem of hyperglycaemia by lowering the absorption and digestibility of starch. It was also investigated that polyphenols inhibit the porcine pancreatic alpha-amylase and rat intestinal alpha-glucosidase. Hydroxyl groups in flavonoids present in finger millet, are responsible for inhibition of starch digestive enzymes (Himanshu, Chauhan, Sonawane, & Arya, 2018). The digestion process can be slowed down with the help of phytochemicals present in finger millet which manage blood sugar levels in diabetic patients (Bisht, Bisht, Gudadhe, Raut, & Dobhal, 2023). According to Jacob et al. (2024) studies, diets containing finger millet helps in reducing blood glucose levels as studies carried out in rats and mice models. A study explored the potential of finger millet-enriched probiotic fermented milk in managing diabetes using streptozotocin-induced diabetic rats. The rats were divided into six groups, each containing six Sprague Dawley rats, with two serving as control groups and the remaining four as test groups. The diabetic groups were administered probiotic fermented milk, finger millet-enriched probiotic fermented milk, finger millet flour, or metformin (a standard diabetes drug) for four weeks. The probiotic strain *Lactobacillus helveticus* MTCC 5463 was used for fermentation and administered orally in finger millet-enriched probiotic fermented milk. The results revealed that the fasting blood glucose levels in diabetic rats treated with finger millet-enriched probiotic fermented milk decreased significantly by 64.26%, outperforming metformin (56.43%), probiotic fermented milk (18.88%), and finger millet flour (47.14%). Additionally, symptoms such as polyphagia (excessive hunger) and polydipsia (excessive thirst) were notably alleviated in diabetic rats after four weeks of treatment. Histological analysis of pancreatic and liver tissues indicated that the finger millet-enriched milk prevented severe damage to acinar cells and reduced inflammation and structural alterations in the liver to a considerable extent. These findings suggest that finger millet-enriched probiotic fermented milk holds promise as a functional food for diabetes management. Its ability to significantly lower blood glucose levels, alleviate diabetes-related symptoms, and protect pancreatic and liver tissues highlights its potential as a natural and effective dietary intervention for individuals with diabetes (Chaudhary & Mudgal, 2020).

Owherua and team deduced that the breakfast cereals produced by blending malted finger millet and watermelon seeds demonstrated a strong inhibitory effect on enzymes alpha-amylase (78.18%) and alpha-glucosidase (56.21%). These enzymes are associated with type-2 diabetes. Furthermore, the combination exhibited a low glycaemic index of 51.25% (Owheruo et al., 2024). According to the available literature, it has been found that finger millet has an intermediate glycaemic index ranging from 55 to 69, which is significantly lower than refined wheat flour, white rice, standard glucose or even white bread. Long-term consumption of finger millet notably reduced the fasting and postprandial blood glucose levels by 12% and 15%. Also, among prediabetic patients, a remarkable decrease in HbA1C levels from 6.65% $\pm$ 0.4% to 5.67% $\pm$ 0.4% was observed (Anitha et al., 2021). Research studies have also indicated that consumption of finger millet accelerated wound healing and postponed cataract development in diabetic rat models (Shobana, Harsha, Platel, Srinivasan, & Malleshi, 2010). Phytic acid, an antinutritional compound present in finger millet has gained popularity as a functional food due to its ability to decrease the rate of carbohydrate digestion, thereby controlling postprandial glucose levels and also aiding in obesity management (Singh et al., 2022). Hence it can be concluded from the reviewed data, that finger millet can be a better alternative in dietary management and the prevention of diabetes. Governments should take steps to promote the beneficial effects of millet consumption for the management of diabetes.

8.2.2 Cardiovascular Health

According to WHO, CVD can be enhanced due to abnormal blood pressure, increased cholesterol levels, depression, diabetes and obesity problems. Various epidemiological studies indicate that these problems are primarily associated with poor dietary habits like frequent intake of foods rich in saturated fats, combined with sedentary lifestyle and inadequate physical activity. Millets contain sterols and pinacosanols, which led to reduction in synthesis of cholesterol. Finger millet contains about 7–12% of protein which includes essential as well as non-essential amino acids. Out of which, lecithin and methionine, found in finger millet, assist in the removal of extra fat from the liver which results in reducing cholesterol levels in the body. In addition, finger millet contains lower concentration of serum triglycerides. Methionine is an important amino acid found in finger millet which is lacking in the low-income group people's diet (Gaikwad 2024). Cholesterol level is also maintained by threonine amino acid which resists fat formation in the body (Bisht, Bisht, Gudadhe, Raut, & Dobhal, 2023). Tryptophan is transformed into metabolites such as serotonin, melatonin, and niacin which help in treatment of CVD, chronic kidney disease, and inflammatory bowel disease (Ashik Somarajan & Morya, 2022). Finger millet offers significant protection against CVDs, such as strokes and heart attacks, by reducing lipid peroxidation, which in turn helps prevent arteriosclerosis. Studies have shown that a multigrain diet incorporating finger millet effectively regulates lipid levels and activates antioxidant mechanisms in high cholesterol-induced rat models. Additionally, the soluble dietary fibres in finger millet reduce the reabsorption of bile acids, which are synthesized from cholesterol, thereby lowering LDL cholesterol levels. Finger millet also contains anti-hypercholesteraemic metabolites like statins, which block the cholesterol synthesis pathway in the liver, further inhibiting cholesterol production (Sawant & Tavhare, 2023). The high fibre and complex carbohydrate content in finger millet contribute to lowering bad cholesterol levels. Ferulic acid, a key component of finger millet, is particularly effective in reducing cholesterol. By including finger millet in the diet, individuals can lower their risk of hypertension, atherosclerosis, and other cardiovascular conditions. Overall, finger millet serves as a natural and effective dietary intervention for promoting heart health and reducing the risk of CVDs (Bisht, Bisht, Gudadhe, Raut, & Dobhal, 2023).

8.2.3 Bone Health

Calcium is an important mineral in finger millet and helps to build strong bones in children as well as old people. Calcium maintains heart health, builds bones, and helps in the synthesis of blood cells. It also helps in avoiding the fracture risk and related disorders like osteoporosis (Bisht, Bisht, Gudadhe, Raut, & Dobhal, 2023). Osteoporosis is a silent disorder due to the loss of bone mass and it is declared as the second major concern for healthcare by WHO. This can be avoided by consuming calcium supplements or naturally available calcium-rich foods. Finger millet is a good source of Phosphorus, vitamin D and B12, and calcium with 350 mg/100 g so it can be used as a food ingredient which helps in reducing the risk of osteoporosis and also leads to bone mass development (Dhurwey & Tripathi, 2024). During Anitha et al. (2021) study, it was found that in children fed with a finger millet-based diet, a significant retention of calcium of about 21–25% was measured whereas in another study of bone turnover marker, Beta CTX is used as a marker for increase in bone resorption and released in blood during bone resorption, 28% and 47% reduction in calcium resorption during peri- and post-menopausal periods in women was found (Seetha Anitha et al., 2021). A study was planned to assess the effect of physical activity in combination with finger millet-based food supplements on the biochemical markers and bone mineral density (BMD) in premenopausal women. After screening 720 women aged between 30 and 40, 150 women who were diagnosed with lower BMD having a t-score between –1 and –2.5 and also lower calcium less than 9.0 mg/dL were selected and divided evenly into control and experimental groups. The experimental group was made to undergo structured physical activity for 5 days every week and were also given finger millet-based sweet balls (laddu) for 3 days per week for 3 months. After 3 months, a significant enhancement in the BMD, biochemical parameters, physical endurance and calcium levels was observed in all the perimenopausal women (Rani & Swaminathan, 2022). In another research experiment by Gayathri and Hemamalini, finger millet-based food formulation with sesame seeds, soyabean, agathi leaves and black gram dal were fed to 60 peri and post-menopausal women. It was observed that bone resorption marker-beta CTX decreased (Gayathri & Hemamalini, 2020). Overall, it can be comprehended that incorporation of finger millet in diet offers an efficient strategy for bone health maintenance from the childhood to old age.

8.2.4 Improved Gut Health

Cereals like wheat, barley and rye contain gluten protein which causes celiac disease, an enteropathic disease which can be avoided only by restricting to gluten-free diet, that is, millet (Kumar et al., 2016). The health benefits of the millets are enhanced by the production of resistant starch which is added to dietary fibre content, highly susceptible to fermentation in the large intestine. Dietary fibre-rich finger millet helps to maintain and improve the intestinal pathways by slowing the digestion process, and provides the bulk which changes the transit time of substances in the body. Adsorption of dietary fibre in small intestinal tract enhances the bile acid excretion which in turn reduces the plasma cholesterol (Devi, Vijayabharathi, Sathyabama, Malleshi, & Priyadarisini, 2014). Low-fat and gluten-free grains of finger millet are responsible for easy digestion in the human body. Phytochemicals in finger millet help in slowing down the digestive process and also help to manage blood sugar levels in case of diabetic patients. Finger millet can also help in reducing glycaemic response which results in lower digestibility and starch absorption (Bisht, Bisht, Gudadhe, Raut, & Dobhal, 2023). Singh, Soni, Noor, and Gupta (2024) studied that finger millets possess various health-beneficial components, out of which soluble and insoluble fibre prevents gastro-intestinal disorders because these components do not break during the process of digestion. Arabinoxylan obtained from finger millet was found to prevent weight gain in high-fat diet-fed male Swiss albino mice. Furthermore, an alteration in response to glucose, serum lipid profile, inflammation and fat accumulation in the liver was observed. The results indicated that arabinoxylan obtained from finger millet could be successfully included as a functional component in diet to manage the high-fat diet-induced obesity (Sarma et al., 2018). Similarly, the polyphenol-rich extracts (PREs) from finger millet were utilized for the development of functional foods for the management of obesity and comorbidities (Khare et al., 2020). In another study, PREs from finger millet helped in the proliferation of good bacteria namely *Bifidobacterium* spp., *Lactobacillus* spp. and short-chain fatty acids in the gut. The polyphenols imparted protective action against Dextran sulphate-induced barrier dysfunction by significantly reducing the inflammation causing cytokines (Maurya et al., 2024).

Millets affect the mood, and behaviour of humans because gut microbiota is attached to the nervous system through the immune system which promotes brain health affected by butyric acid produced by fermentation. A millet-rich diet helps to improve gut health and nutritional value. Probiotics like *Bifidobacterium* and *Lactobacillus*, and Butyrate producers such as *Faecalibacterium* and *Eubacterium* help in inhibiting pathogens which regulates toxin production and reduces inflammation in the gut as shown in Fig. 8.2 (Singh et al., 2022). Prebiotic-resistant starch led to a rise in many beneficial microflora including *Bifidobacterium*, *Lactobacillus*, and *Akkermansia*

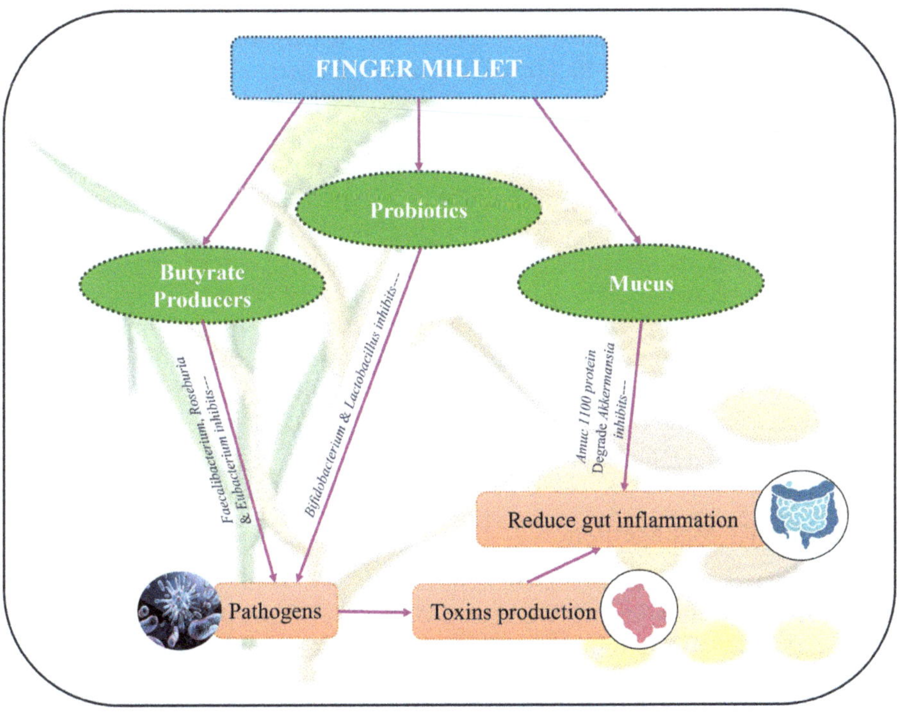

FIGURE 8.2
Effect of finger millet on gut health.

and also resistant to enzymatic digestion (Jacob et al., 2024). When finger millet is subjected to fermentation, it increases the number of *Pediococcus* and reduces the pathogenic organisms. It was found that regular consumption of finger millet in the existing diet led to anti-diabetic and anti-obesity effects. Finger millet contains various compounds which help in the growth of gut microbiota such as butyrate-producing *Roseburia* and *Faecalibacterium* which helps in the reduction of inflammation by inhibitory action on pathogens (Singh et al., 2022).

8.2.5 Rapid Wound Healing and Production of Nerve Growth Factor

Himanshu, Chauhan, Sonawane, and Arya (2018) investigated that skin wounds can be treated within 2 weeks with prior feeding of a finger millet diet. It was found that the healing process is enhanced with the increase in the concentration of wounds in hyperglycaemic rats by giving them finger-millet-rich diets. Along with wound healing, it was found that the index of oxidative stress and TBA (thiobarbituric acid) reactive substances was enhanced in wound tissues as compared to that of normal tissue. Also by feeding a finger millet diet to rats, an impaired production of nerve growth factor (NGF) was enhanced which can be confirmed by immunocytochemical studies. A delay in the wound healing process is caused by increasing oxidative stress markers results in reduced antioxidant activity but it was found that antioxidant activity is increased significantly by feeding finger millet in diabetic rats. The wound healing process can be understood by examining the process of tissue repairment that can be determined by inflammation. Increased levels of reactive oxygen species may interrupt the wound healing process mainly in the case of diabetic patients and age-associated diseases (Devi, Vijayabharathi, Sathyabama, Malleshi, & Priyadarisini, 2014). Wound healing is enhanced in diabetic rats when they fed with finger millet through the activities of antioxidant enzymes, enhancement of nerve growth factor, increased production of collagen and activation of cells, etc. (Jacob et al., 2024). Udeh and colleagues deduced that polyphenols present in finger millet when incorporated in wound healing films showcased an improvement in diabetic foot ulcers. Also, the calcium in finger millet aids in diabetic wound healing (Udeh, Duodu, & Jideani, (2017)

8.2.6 Obesity Management

Obesity is a chronic complex disease that is defined by the accumulation of excess fat. According to a report by WHO, 1 in 8 people in the world are obese. By 2022, approximately 2.5 billion adults aged ≥18 years were classified as overweight. It also raised concerns about childhood obesity where 37 million children under 5 years were designated as overweight (WHO, 2024). One of the effective strategies to overcome the issue of obesity is to replace refined grains with whole grains like millet. Millets are not only cheap and easily available but also offer augmented nutritional benefits with weight management. Finger millet contains various amino acids out of which, tryptophan helps in the reduction of hunger and maintain a healthy weight as it digests slowly, so it helps to avoid too eating. Also, fibres in millet make you feel full which also helps in a controlled diet (Bisht, Bisht, Gudadhe, Raut, & Dobhal, 2023). According to the study of (Ashik Somarajan & Morya, 2022), phenols present in the seed coat of finger millet with varying molecular weights like tannins and anthocyanins. These phenolic compounds were found to be effective against carbohydrate-hydrolysing enzyme inhibitors. Finger millet contains a higher amount of dietary fibre than other millet as it has five layered seed coats which provide more satiety by consuming less food (Kaur et al., 2024). Krishnan and team assessed the efficacy of finger millet and pearl millet-based breakfast cereal for anti-obesity characteristics. The formulations exhibited excellent antioxidant activity and enzyme (lipase and amylase) inhibition activity suggesting potential effectiveness against obesity (Krishnan et al., 2024). A study was planned by Ibrahem and Sayed to evaluate the effect of the replacement of wheat flour with finger millet flour in obese rats. It was contemplated that supplementation of finger millet flour led to a reduction in weight gain; feed efficiency ratio and body mass index. A decrement in triglyceride levels and cholesterol values was also reported (Ibrahem & Sayed, 2023).

8.3 Therapeutic Applications of Finger Millet

Finger millet has been an essential component of traditional medicine for centuries, deeply embedded in the cultural practices of India, Africa, and Southeast Asia. It is commonly consumed in the form of porridges, breads, and fermented beverages, believed to enhance physical strength, boost immunity, and promote digestive health. In Ayurvedic medicine, finger millet is used to balance doshas and is recommended for its cooling properties, aiding in the treatment of fevers and inflammatory conditions. Its rich calcium and iron content are utilized to prevent bone-related disorders and anaemia, particularly in pregnant and lactating women. In Africa, finger millet is used in postpartum recovery and to manage diabetes, attributed to its low glycaemic index and high fibre content. Decoctions prepared from finger millet grains are also employed to alleviate headaches and promote restful sleep, reflecting its holistic role in regional remedies (Gaikwad, 2024).

8.3.1 Cancer Prevention

During the last years, cases of cancer patients increased subsequently in India which can be treated by the use of traditional medicine including millet consumption as it contains many beneficial nutraceutical components like phytochemicals and antioxidants which help in reducing the cancer risk by inhibiting the cancer cells because these compounds help in terminating free radical and oxygen species (Kumar et al., 2016). Finger millet contains phenolic compounds and antioxidants, which exhibit anticarcinogenic properties. Polyphenols, particularly ferulic acid and catechins, have been shown to inhibit the proliferation of cancer cells. These compounds act by neutralizing free radicals and reducing oxidative stress, thereby preventing DNA damage and the initiation of cancer. Phenolic compounds extracted from finger millet exhibited cytotoxic potential against breast cancer cell lines (Kuruburu, Bovilla, Naaz, Leihang, & Madhunapantula, 2022). Fermented finger millet formulation with *Levilactobacillus brevis* exhibited antiproliferative and anti-migratory effects against MDA-MB-231 and HCT116 cancer cell lines (Mahanta et al., 2024). Sathiya Narayanan et al. (2024) reported that phytochemicals from finger millet showcased potency against matrix metalloproteinase-I, responsible for oral cancer, by blocking its active site. Ethanolic and methanolic extracts of Sri Lankan finger millet varieties displayed inhibitory action against glutathione S-transferase and β-glucuronidase enzymes. Hence, these extracts can be utilized in the treatment of liver and colon cancer. In addition, these extracts were found to be cytotoxic against breast and lung cancer cell lines also (Jayawardana et al., 2020). Studies on animal models reveal that finger millet extracts reduce tumour growth and mitigate oxidative stress, suggesting a protective role against colon, liver, and breast cancer (Zhang et al., 2021). Additionally, the dietary fibre present in finger millet facilitates the removal of carcinogens from the gastrointestinal tract, contributing to colon cancer prevention. Regular consumption of finger millet products has been linked to lower incidences of gastrointestinal malignancies, reinforcing its role as a functional food in cancer prevention (Devi, Vijayabharathi, Sathyabama, Malleshi, & Priyadarisini, 2014). Along with this, ferulic acid also helps to block carcinogenesis in the colon and tongue of rats. Vitamin B17 (nitroliside) is also present in finger millet which is used to treat cancer without showing any side effects (Kumar et al., 2016). Hence it can be concluded from the above discussion that bioactive compounds from finger millet are promising functional foods for cancer prevention and treatment.

8.3.2 Neuroprotective Effects

Finger millet is rich in polyphenols, flavonoids, and antioxidants, which contribute to its neuroprotective properties. Key compounds such as ferulic acid, catechins, quercetin, tannins, and phenolic acids play vital roles in scavenging free radicals, inhibiting lipid peroxidation, and protecting neural cells. Additionally, the dietary fibre and essential minerals present in finger millet contribute to overall brain health and reduce inflammation, supporting long-term neural function.

Incorporating finger millet into the diet can provide long-term neuroprotection. Whole grain flour can be used in traditional porridge and bread, while millet-based snacks rich in polyphenols offer convenient options for neuroprotection. Functional foods fortified with additional neuroprotective compounds are also emerging as effective dietary interventions. Regular consumption of finger millet in various forms can help mitigate the effects of oxidative stress and support brain health across different age groups (Jayawardana et al., 2021). The neuroprotective effects of finger millet polyphenols (FMP) are mediated through various mechanisms. These compounds reduce oxidative stress by neutralizing reactive oxygen species and enhancing antioxidant enzyme activity, including superoxide dismutase and catalase. Finger millet also exhibits anti-inflammatory properties by downregulating pro-inflammatory cytokines, thereby reducing neuroinflammation. Furthermore, FMP protect mitochondrial function, preventing neuronal apoptosis and enhancing energy production. By regulating lipid metabolism and lowering lipid peroxidation, finger millet safeguards against high-fat diet-induced brain damage. Finger millet's neuroprotective properties make it a promising candidate for managing neurodegenerative diseases such as Alzheimer's disease, Parkinson's disease, and stroke recovery. In Alzheimer's disease, FMP reduce amyloid-beta plaque formation and oxidative damage. In Parkinson's disease, they help preserve dopamine levels and protect against neuronal degeneration. Finger millet also enhances neurogenesis and reduces infarct size, aiding in stroke recovery and improving overall brain resilience (Li et al., 2020).

8.3.3 Anti-Aging Potential

Finger millet is abundant in antioxidants, including phenolic acids, flavonoids, and tannins. These compounds neutralize free radicals, reducing oxidative stress—a key contributor to cellular aging and various degenerative diseases. The seed coat of finger millet is especially rich in these antioxidants, making whole-grain consumption particularly beneficial (Devi, Vijayabharathi, Sathyabama, Malleshi, & Priyadarisini, 2014).

Finger millet contains essential amino acids, notably lysine, which is crucial for collagen synthesis. Collagen is a structural protein vital for maintaining skin firmness and elasticity. Adequate collagen production helps prevent wrinkles and sagging skin, common signs of aging. Incorporating finger millet into the diet can support the body's natural collagen production, promoting youthful skin. Lysine, present in finger millet, is responsible for the reduction of risk of CVD, cancer as well as aging in human beings. The study of riboflavin has been widely carried out for its antioxidant, anti-aging, and anti-nociceptive properties (Ashik Somarajan & Morya, 2022).

Finger millet is a rich source of minerals such as calcium, iron, and zinc. Calcium is essential for skin cell renewal, while zinc plays a role in maintaining skin integrity and repairing damaged tissues. Iron supports proper oxygenation of skin cells, contributing to a healthy complexion. The presence of these minerals in finger millet aids in maintaining skin health and delaying the onset of age-related skin issues.

The high dietary fibre content in finger millet facilitates efficient digestion and detoxification. A healthy digestive system ensures the elimination of toxins, which can otherwise contribute to premature aging. Additionally, dietary fibre helps maintain a balanced gut microbiome, which has been linked to improved skin health and reduced inflammation (Ashik Somarajan & Morya, 2022).

Maintaining stable blood sugar levels is crucial for preventing glycation, a process where sugar molecules bind to proteins, leading to the formation of advanced glycation end-products (AGEs). AGEs contribute to skin aging by damaging collagen and elastin fibres. Finger millet has a low glycaemic index, promoting gradual blood sugar release and reducing the risk of glycation, thereby preserving skin elasticity and preventing premature aging (Bisht, Bisht, Gudadhe, Raut, & Dobhal, 2023).

8.3.4 Inhibition of Cataract Development

Himanshu, Chauhan, Sonawane, and Arya (2018) investigated that the problems related to cataracts and retinopathy are prominent in Western countries as a major cause of blindness. Cataract is responsible for 80% of the cases of prevailing blindness. An accumulation of sorbitol leads to diabetes and induces cataracts, which is regulated by the enzyme aldose reductase (AR). The mechanism involved in this cataract was mainly the binding of glucose to protein which forms non-enzymatic glycation induced during diabetes leading to enzyme-mediated sugar-induced cataract. It was also studied that polyphenols in finger millet such as protocatechuic, gallic, vanillic, syringic, and ferulic acids, etc., reduce the risk of cataract effectively. The activity of AR was significantly reduced in the presence of O-methyl group. According to Jacob et al. (2024) studies, it was found that phenols in finger millet delay cataract development in rats by reducing the activity of AR, which also lowers down the glycosylated Hb level in blood. The accumulation of sorbitol, catalysed by the enzyme AR, plays a crucial role in the onset of diabetes-induced cataracts. Finger millet was studied for its ability to inhibit AR due to the presence of FMP, which possess strong anti-diabetic and antioxidant properties. The phenolic compounds in FMP, such as gallic acid, p-hydroxybenzoic acid, protocatechuic acid, p-coumaric acid, syringic acid, vanillic acid, trans-cinnamic acid, ferulic acid, and quercetin, demonstrated significant effectiveness in preventing cataract formation in the eye lens. Furthermore, polyphenols derived from the seed coat of finger millet inhibited **AR** through reversible non-competitive inhibition. These results provide compelling evidence of FMP's potential in mitigating cataractogenesis in humans (Nayak & Satish, 2024). Accumulation of sorbitol in human body can be avoided by inhibiting the AR which reduces the chances of cataract disorders due to diabetes (Devi, Vijayabharathi, Sathyabama, Malleshi, & Priyadarisini, 2014).

8.3.5 Natural Immunomodulator

The immunostimulatory potential of purified arabinoxylans extracted from finger millet bran was thoroughly evaluated through experiments involving murine lymphocytes and peritoneal exudate macrophages. The results demonstrated that the arabinoxylans significantly enhanced macrophage phagocytosis, showing a notable twofold increase in activity. Further analysis indicated that the immunostimulatory effects of these purified arabinoxylans were directly linked to their ferulic acid content, suggesting that higher levels of ferulic acid contribute to greater immune-boosting properties. These findings highlight the promising role of purified arabinoxylans from finger millet bran as a potent and natural immunomodulatory agent, which could be further explored for therapeutic or functional food applications (Savitha Prashanth et al., 2014).

8.3.6 Colorectal Carcinoma and Breast Cancer

Finger millet is a nutrient-dense grain, known for its high content of dietary fibre, tannins, phenolic compounds, and calcium. The study investigated the effects of phenolic compounds derived from finger millet on colorectal carcinoma and breast cancer cell lines (Siegel, Miller, Fuchs, & Jemal, 2021). It was observed that the phenolic fractions displayed significant ferric ion reduction capacity and 2,2-Diphenyl-1-picrylhydrazyl (DPPH) radical scavenging activity, underscoring their potent antioxidant properties (Zhang et al., 2021). Notably, the free phenolic compounds from finger millet demonstrated superior cytotoxic effects against breast cancer cell lines compared to the bound phenolic compounds, indicating their enhanced ability to inhibit cancer cell growth. These findings highlight the efficacy of both free and bound phenolic compounds in finger millet, showcasing their antioxidant potential and their ability to modulate the proliferation of breast and colorectal cancer cells. The study emphasizes the promising role of finger millet phenolic compounds as natural agents with anticancer and antioxidant properties, offering potential therapeutic benefits in combating cancer and oxidative stress-related conditions (Kuruburu, Bovilla, Naaz, Leihang, & Madhunapantula, 2022).

8.3.7 Anti-Hypertensive Activities

According to (Park et al., 2021) study, about 80% patients of with hypertension were not familiar with their problem. A major cause of hypertension is reactive oxygen species resulting in increased blood pressure due to changes in diets and living environments. Hypertension is mainly caused by ACE (angiotensin converting enzyme), responsible for the renin-angiotensin-aldosterone system by producing Angiotensin II from the conversion of Angiotensin I. Hypertension-like situations can be mediated by synthetic ACE inhibitors like captopril and enalapril but the major issues with these inhibitors are their side effects on health like cough and skin problems. To avoid such issues, a well-planned diet is necessary to consume which helps in controlling the blood pressure in the human body. Millets contain a large amount of anti-hypertensive components including fibre content, protein minerals, etc. Ethanolic extract of finger millet helps in controlling the rennin angiotensin mechanism which is anti-hypertensive (Jacob et al., 2024). These extracts help in maintaining ACE levels in the body by enhancing the antioxidant activity which regulates the production of reactive oxygen species. Antioxidant capacity was assessed by thiobarbituric acid reactive substance (TBARS) concentration and superoxide dismutase (SOD) activity which was significantly reduced in rats when they were fed with finger millet in their diet (Park et al., 2021).

8.3.8 Antioxidant Effects

According to the studies (Kumar et al., 2016), it was studied that reactive oxygen and nitrogen species lead to oxidation of cellular molecules which are responsible for various chronic disorders such as diabetes, cardiac problems, cancer, etc. The major phenolic component in finger millet is protocatechuic acid which is present in free form and also contains benzoic acid derivatives such as gallic acid, vanillic acid, syringic acid, and p-hydroxybenzoic acid. Phenolic content in finger millet including gallic acid, is responsible for antioxidant effect as gallic acid is a trihydroxybenzoic acid playing a key role as an astringent, a human xenobiotic metabolite (Ashik Somarajan & Morya, 2022). Vitamins available from finger millet help to show antioxidant activities. Phytates and polyphenols exhibiting antioxidant properties, also play an important role in maintaining good health, delay of the aging process, and preventing metabolic disorders. Finger millet can also help in reducing glycaemic response which results in lower digestibility and starch absorption (Bisht, Bisht, Gudadhe, Raut, & Dobhal, 2023). Jayawardana et al. (2021) also concluded that Ravi, Rawana, and Oshadha varieties of finger millet are also responsible for scavenging free radicals which showed high antioxidant activities due to the presence of flavonoids in finger millet. Antioxidant activities of finger millet have been checked against scavenging DPPH, ABTS cation, and oxygen radicals. It has also been studied that a continuous intake of finger millet enriched diet may aid in reducing the risk of oxidative stress-associated diseases. The demand for antioxidants is increasing to reduce the excessive oxidation in the human body that leads to the growth of cancer cells and aging. Antioxidants work as reducing agents, lipid stabilizers, and metal chelators and their activity depends on their ability to donate hydrogen atoms to electron-deficient free radicals by producing a stable radical that is less reactive (Devi, Vijayabharathi,

Sathyabama, Malleshi, & Priyadarisini, 2014). It was studied that a diet containing 55% finger millet showed an increase in the antioxidant activity of enzymes such as catalase, glutathione reductase peroxidase, etc., in rats model (Kumar et al., 2016). According to the studies by Kumar et al. (2016), the phenolic content of finger millet is reduced during the processing of finger millet such as germination, decortication, thermal processing fermentation, etc.

8.3.9 Anti-Inflammatory Effects

The immune system creates a defensive mechanism by producing inflammation to protect against pathogens, damaged cells, and allergens but in excessive inflammation, it can lead to acute and chronic diseases. Different types of inflammatory disorders are caused by the enzyme Arachidonate 5-lipoxygenase (A5-LOX) which involves the production of leucotrienes catalysing from arachidonic acid. It was found during the studies of anti-inflammatory properties of different varieties of finger millet such as Ravi, Rawana and Oshadha, etc., that methanolic extracts of these varieties showed higher inhibitory action against A5-LOX enzyme as compared to that of ethanolic extracts of these varieties. Inflammatory disorders can be regulated by Xanthine oxidase inhibitors which is responsible for the production of uric acid from xanthine. It was also found that hyaluronidase helps in hydrolysis of mucopolysaccharides such as hyaluronic acid. The highest inhibitory activity against XO enzymes had been found in Oshadha varieties of finger millet. Hyaluronidase degrades hyaluronic acid which led to a reduction in molecular weight and its amount. Inhibitory activities of both methanolic and ethanolic extract of finger millet have been studied effective against hyaluronidase (Jayawardana et al., 2021). Finger millet contains soluble fibres that can be useful in assisting lubrication and smoothening effect on the inflamed digestive tract. Polyphenol found in finger millet also helps in reducing peptic inflammation and shows anti-ulcerative effects (Kumar et al., 2016)

8.3.10 Antimicrobial Properties of Finger Millet

Finger millet contains various components which show antimicrobial activity against various microflora. Phenolic extract from the seed coat of finger millet are active against *Bacillus,* and *Aspergillus* species. Extract from fermented finger millet shows inhibition against the microflora namely, *salmonella* and *E.coli.* Phenol extract from the germinated and ungerminated seed of millet shows an action against *Bacillus, Staphylococcus, Yersinia, E.coli,* and *Listeria* species. Higher antibacterial and antifungal activity has been shown by the acidic methanolic extract from the seed coat of finger millet as compared to that of whole millet flour because of high polyphenolic content in the seed coat of finger millet. Polyphenols in finger millet showed antifungal activity by the formation of irreversible complexes with nucleophilic amino acids and oxidation of cell membranes and components (Devi, Vijayabharathi, Sathyabama, Malleshi, & Priyadarisini, 2014). The study of (Ashik Somarajan & Morya, 2022) found that the antimicrobial activity of finger millet can be determined by observing the beta carotene autoxidation in the emulsion of linolenic acid. The polyphenols extracted from finger millet demonstrated remarkable proliferation inhibitory activities against a wide spectrum of pathogenic bacteria, including *Escherichia coli, Bacillus cereus, Staphylococcus aureus, Listeria monocytogenes, Streptococcus pyogenes, Proteus mirabilis, Pseudomonas aeruginosa, Serratia marcescens, Klebsiella pneumoniae,* and *Yersinia enterocolitica.* Finger millet grains can be protected against fungal attack due to the presence of tannin in the outer layer or seed coat of grains (Abioye et al., 2022). Among the various polyphenols tested, quercetin stood out due to its broad-spectrum antibacterial properties, effectively inhibiting the growth of all the bacterial strains examined. On the other hand, other polyphenolic compounds such as gallic acid, ferulic acid, proto-catechuic acid, and p-hydroxy benzoic acid exhibited more selective antibacterial activity, primarily targeting specific pathogens. The significant inhibitory effects of quercetin on these pathogenic bacteria highlight the potent antimicrobial potential of FMP, suggesting their valuable role as natural nutraceuticals. The polyphenols in finger millet not only possess antimicrobial properties but also serve as a rich source of natural antioxidants. These antioxidants can play a crucial role in mitigating oxidative stress, which is often linked to the development of various chronic diseases. By incorporating FMP into diets or therapeutic formulations, it may be possible to reduce the risk of oxidative damage and associated health complications. Overall FMP hold great promise as a natural, multifunctional ingredient for promoting health and preventing diseases caused by both microbial infections and oxidative deterioration (Nayak & Satish, 2024).

8.4 Conclusion

With the increase in nutritional awareness, it becomes a necessity for the different food industries to look out towards the development of new food products with high nutritional content which impart various health benefits. Cereals and millet serve as fundamental food sources globally, including in India and various foreign countries. Among the diverse varieties of millet, finger millet (*E. coracana*), commonly known as ragi, holds significant importance as a staple food in regions of eastern and central Africa, as well as in India. Finger millet is gaining recognition for its substantial health benefits, attributed to its rich content of essential vitamins and minerals. Notably, it is a superior source of calcium, iron, dietary fibre, and antioxidants, which contribute to its role in promoting good health. Finger millet is also known for its ease of digestion, making it suitable for individuals across different age groups. In recent decades, there has been a notable increase in the consumption of finger millet, reflecting a shift in dietary preferences towards more nutritious and health-promoting foods. This trend is further supported by the crop's adaptability to various soil types and climatic conditions, facilitating its cultivation across multiple regions worldwide. India, for instance, dominates global millet production, accounting for approximately 42% of the total output. The development of products made from composite flours, incorporating finger millet, has resulted in nutritionally superior alternatives compared to traditional options. These products can be effectively utilized in supplementary feeding programs, aiming to enhance nutritional intake among various populations. To maximize the benefits of finger millet, it is imperative to educate communities about its nutritional value and health advantages. Such awareness can lead to increased acceptance and incorporation of finger millet-based foods into daily diets. Consequently, steps should be taken to boost the production and productivity of finger millet through the adoption of improved agricultural technologies and practices. This chapter provides a detailed summary of finger millet including its nutraceuticals and therapeutic properties, contribution to food security and the promotion of public health on a broader scale.

References

Abioye, V. F., Babarinde, G. O., Ogunlakin, G. O., Adejuyitan, J. A., Olatunde, S. J., & Abioye, A. O. (2022). Varietal and processing influence on nutritional and phytochemical properties of finger millet: A review. *Heliyon, 8*(12), e12310. https://doi.org/10.1016/j.heliyon.2022.e12310

Anitha, S., Givens, D. I., Botha, R., Kane-Potaka, J., Sulaiman, N. L. B., Tsusaka, T. W., Subramaniam, K., Rajendran, A., Parasannanavar, D. J., & Bhandari, R. K. (2021). Calcium from finger millet—A systematic review and meta-analysis on calcium retention, bone resorption, and in vitro bioavailability. *Sustainability, 13*(16), 8677. https://doi.org/10.3390/su13168677

Ashik Somarajan K, A., & Morya, S. (2022). A review on finger millet properties, processing, health benefits, and applications. *The Pharma Innovation, 11*, 3388–3395. https://doi.org/10.22271/tpi.2022.v11.i7aq.14538

Bhujle, R. R., Nayak, N., Gowda, N. N., Pandiselvam, R., & Sunil, C. K. (2024). A comprehensive review on influence of millet processing on carbohydrate-digesting enzyme inhibitors and implications for diabetes management. *Critical reviews in Biotechnology, 45*(4), 743–765. https://doi.org/10.1080/07388551.2024.2423153

Bisht, K., Bisht, K., Gudadhe, N. N., Raut, A. A., & Dobhal, N. (2023). Nutritional composition, health benefits, production, processing and marketing of finger millet. *Indian Journal of Fertilisers, 19*(10), 1036–1046.

Chand, R. D., Rajnish, R., Chandra, H., Dwivedi, R. S., & Singh, R. P. (2024). Enhancing health through telemedicine and Ayurvedic resources: A sustainable model. *International Journal of Plant and Environment, 10*(03), 131–137. https://doi.org/10.18811/ijpen.v10i03.15

Chaudhary, J. K., & Mudgal, S. (2020). Antidiabetic and hypolipidaemic action of finger millet (*Eleusine coracana*)-enriched probiotic fermented milk: An in vivo rat study. *Food Technology and Biotechnology, 58*(2), 192–202. https://doi.org/10.17113/ftb.58.02.20.6308

Choudhury, S. S., & Chaudhary, G., (2024). Impact of iron rich millet biscuits in alleviating anaemia among adolescent girls of Samastipur district. *Asian Journal of Dairy and Food Research, 43*(1), 111–115. https://doi.org/10.18805/ajdfr.DR-2071

Devi, P. B., Vijayabharathi, R., Sathyabama, S., Malleshi, N. G., & Priyadarisini, V. B. (2014). Health benefits of finger millet (*Eleusine coracana* L.) polyphenols and dietary fiber: A review. *Journal of Food Science and Technology, 51,* 1021–1040. https://doi.org/10.1007/s13197-011-0584-9

Dhurwey, R., & Tripathi, R. (2024). Assessing the health benefits of finger millet – A review. *International Journal of Creative Research Thoughts, 12*(9), 282–287.

Dube, S., Varade, S., & Choudhari, V. (2024). Utility of Finger millet (ragi) in postmenopausal osteoporosis: A Review. World J Pharm Res, *13,* 272–279. 10.20959/wjpr20246-31637

Gaikwad, V. (2024). Nutritional significance of finger millet and its potential for using in functional products. *Foods and Raw Materials, 12*(1), 110–123. https://doi.org/10.21603/2308-4057-2023-1-547

Gayathri, G., & Hemamalini, A. J. (2020). Plant based Indigenous dietary calcium supplementation on bone turnover markers among Peri and postmenopausal women: A randomised controlled trial. *Indian Journal of Community Health, 32*(4), 705–712. https://doi.org/10.47203/IJCH.2020.v32i04.016

Himanshu, K., Chauhan, M., Sonawane, S. K., & Arya, S. S. (2018). Nutritional and nutraceutical properties of millets: A review. *Clinical Journal of Nutrition and Dietetics, 1*(1), 1–10.

Ibrahem, E. S., & Sayed, R. S. (2023). Characteristics of finger millet (*Eleusine coracana* L.) flour and its effect on obese rats: A biological and applied study. *Egyptian Journal of Food Science, 51*(1), 113–125. https://doi.org/10.21608/ejfs.2023.208211.1164

Jacob, J., Krishnan, V., Antony, C., Bhavyasri, M., Aruna, C., Mishra, K., Nepolean, T., Satyavathi, C. T., & Visarada, K. B. R. S. (2024). The nutrition and therapeutic potential of millets: An updated narrative review. *Frontiers in Nutrition, 11,* 1346869. https://doi.org/10.3389/fnut.2024.1346869

Jayawardana, S. A. S., Samarasekera, J. K. R. R., Hettiarachchi, G. H. C. M., Gooneratne, J., Choudhary, M. I., & Jabeen, A. (2021). Anti-inflammatory and antioxidant properties of finger millet (*Eleusine coracana* (L.) Gaertn.) Varieties cultivated in Sri Lanka. *BioMed Research International, 2021*(1), 1–10. https://doi.org/10.1155/2021/7744961

Jayawardana, S. A. S., Samarasekera, J. K. R. R., Hettiarachchi, G. H. C. M., Gooneratne, J., Choudhary, M. I., Imad, R., & Naz, A. (2020). Glutathione S-transferase and β-glucuronidase enzymes inhibitory and cytotoxic activities of ethanolic and methanolic extracts of Sri Lankan finger millet (*Eleusine coracana* (L.) Gaertn.) varieties. *South Asian Research Journal of Natural Products, 3*(2), 1–9.

Kaur, S., Kumari, A., Seem, K., Sekhon, G., Kumar, D., Verma, S., Singh, N., Kumar, A., Kumar, M., Jaiswal, S., Bhardwaj, R., Singh, B. K., & Riar, A. (2024). Finger millet (*Eleusine coracana* L.): From staple to superfood—A comprehensive review on nutritional, bioactive, industrial, and climate resilience potential. *Planta, 260*(3), 75. https://doi.org/10.1007/s00425-024-04502-2

Kavitha, U., Nihar, S. S. S., & Kader, M. M. P. M. A. (2024). The therapeutic potential of millets: A comprehensive review. *Journal of Research in Siddha Medicine,7*(Supplement 1), S29–S35. https://doi.org/10.4103/jrsm.jrsm_15_23

Khare, P., Maurya, R., Bhatia, R., Mangal, P., Singh, J., Podili, K., Bishnoi, M., & Kondepudi, K. K. (2020). Polyphenol rich extracts of finger millet and kodo millet ameliorate high fat diet-induced metabolic alterations. *Food & Function, 11*(11), 9833–9847. https://doi.org/10.1039/D0FO01643H

Krishnan, H., Kaushik, D., Kumar, M., Oz, E., Brennan, C., Oz, F., Proestos, C., Hashem, A., Ibrahimova, U., & Abd Allah, E. F. (2024). Assessing the efficacy of pearl millet and finger millet enriched breakfast cereals to combat obesity. *Scientific Reports, 14*(1), 27752.

Kumar, A., Metwal, M., Kaur, S., Gupta, A. K., Puranik, S., Singh, S., Singh, M., Gupta, S., Babu, B. K., Sood, S., & Yadav, R. (2016). Nutraceutical value of finger millet [*Eleusine coracana* (L.) Gaertn.], And their improvement using omics approaches. *Frontiers in Plant Science, 7*. https://doi.org/10.3389/fpls.2016.00934

Kuruburu, M. G., Bovilla, V. R., Naaz, R., Leihang, Z., & Madhunapantula, S. V. (2022). Variations in the anticancer activity of free and bound phenolics of finger millet (*Eleusine coracana* (L) Gaertn; variety KMR-301) seeds. *Phytomedicine Plus, 2*(2), 100276. https://doi.org/10.1016/j.phyplu.2022.100276

Li, S., Xian, F., Guan, X., Huang, K., Yu, W., & Liu, D. (2020). Neural protective effects of millet and millet polyphenols on high-fat diet-induced oxidative stress in the brain. *Plant Foods for Human Nutrition, 75*, 208–214.

Mahanta, S. K., Nayak, P. P., Muduli, K., Elangovan, S., Paramasivan, S. S., Mallick, P. K., Mohapatra, S. K., & Panda, S. K. (2024). Optimisation of *Levilactobacillus brevis*-fermented finger millet (*Eleusine coracana*) and evaluation of its effects on cancer cells (HCT116 and MDA-MB-231). *Methods, 229*, 30–40. https://doi.org/10.1016/j.ymeth.2024.06.002

Mansoria, P., & Singh, S. B. (2023). Unlocking the therapeutic potential of millets: A path to diabetes control. *Journal of Ayurveda and Integrated Medical Sciences, 8*(6), 152–157. https://doi.org/10.21760/jaims.8.6.25

Maurya, R., Shandilya, C., Raina, N., Rawat, A., Devi, K., Bhadada, S. K., Bishnoi, M., & Kondepudi, K. K. (2024). Effect of millet polyphenol-rich extracts on human fecal bacteria and dextran sulfate sodium-induced human intestinal epithelial cell derangements. *Food Biotechnology, 38*(1), 42–63. https://doi.org/10.1080/08905436.2024.2305978

Meti, S. C., & Rawat, S. (2024). Finger Millet: Biology, Functional Potential and Sustainable Utilization. In *Millets: The Multi-Cereal Paradigm for Food Sustainability* (pp. 75–88): Springer.

Nayak, R., & Satish, S. (2024). The superfood finger millet (*Eleusine coracana*) and its therapeutic activities – A review. *International Journal of Research and Review, 11*, 372–379. https://doi.org/10.52403/ijrr.20240543

Owheruo, J. O., Edo, G. I., Ifesan, B. O., Bolade, M. K., Origbemisoye, B. A., Akpoghelie, P. O., & Agbo, J. J. J. V. (2024). Evaluation of nutraceutical property of extruded breakfast cereal produced from blends of malted finger millet (*Eleusine coracana*) and watermelon (*Citrullus lanatus*) seed flour. *Vegetos, 37*(6), 2347–2361. https://doi.org/10.1007/s42535-023-00728-9

Park, S. Y., Jeong, E. W., Yang, Y. S., Kim, H. J., Go, G. W., & Lee, H. G. (2021). Finger millet ethanol extracts prevent hypertension by inhibiting the angiotensin-converting enzyme level and enhancing the antioxidant capacity in spontaneously hypertensive rats. *Antioxidants (Basel), 10*(11). https://doi.org/10.3390/antiox10111766

Parveen, S., Kumar, R. S., Rathi, A., Nayeem, A., Qais, F. A., Ahmad, F., & Ur Rehman, S. (2025). In-silico investigation of *Eleusine coracana* phytocompounds as prostate cancer therapeutics targeting heat shock protein 90 alpha family protein. *Journal of Computational Biophysics and Chemistry, 24*(7), 933–948. https://doi.org/10.1142/S2737416525500036

Patil, P., Singh, S., & Patel, P. (2023). Functional properties and health benefits of finger millet (*Eleusine coracana* L.): A review. *The Journal of Phytopharmacology, 12*, 196–202. https://doi.org/10.31254/phyto.2023.12308

Premachandran, K., Wilson Alphonse, C. R., & Soundharapandiyan, N. (2023). Nourishing the cognition with millets: A comprehensive review of their nutritional impact and potential as cognitive enhancers. *Molecular Nutrition & Food Research, 67*(24), 2300450. https://doi.org/10.1002/mnfr.202300450

Rani, G. S., & Swaminathan, A. (2022). Effectiveness of physical activity and diet supplementation on body mass index and bone mineral density among premenopausal women. *Journal of Pharmaceutical Negative Results, 13*(7), 4403–4411. https://doi.org/10.47750/pnr.2022.13.S07.552

Rathore, K. S., & Sharma, A. K. (2012). Nutraceutical importance of finger millet (*Eleusine coracana*) for improved human health. *The European Journal of Plant Science and Biotechnology, 6*(2), 91–95.

Sarma, S. M., Singh, D. P., Singh, P., Khare, P., Mangal, P., Singh, S., Bijalwan, V., Kaur, J., Mantri, S., Boparai, R. K., Mazumdar, K., Bishnoi, M., Bhutani, K. K., & Kondepudi, K. K. (2018). Finger millet arabinoxylan protects mice from high-fat diet induced lipid derangements, inflammation, endotoxemia and gut bacterial dysbiosis. *International Journal of Biological Macromolecules, 106*, 994–1003. https://doi.org/10.1016/j.ijbiomac.2017.08.100

Sathiya Narayanan, S., Kanishka, R., Sowmiyan, M. N., Nithish, V., Kumar, V. R., Mary, A. V., & Kesavan, R. (2024). Targeting MMP1 in oral cancer: In silico analysis of phytochemicals from finger millet as potential inhibitors. *International Journal of Applied Dental Sciences, 10*(4), 264–268. https://doi.org/10.22271/oral.2024.v10.i4d.2074

Savitha Prashanth, M. R., Shruthi, R. R., & Muralikrishna, G. (2014). Immunomodulatory activity of purified arabinoxylans from finger millet (Eleusine coracana, v. Indaf 15) bran. *Journal of Food Science and Technology, 52*(9), 6049–6054.

Sawant, P. D., & Tavhare, S. D. (2023). A review on the use and scope of finger millet (Ragi) in prevention and management of non-communicable diseases. *Journal of Drug Research in Ayurvedic Sciences, 8*(Suppl 1), S55–S63.

Shobana, S., Harsha, M. R., Platel, K., Srinivasan, K., & Malleshi, N. G. (2010). Amelioration of hyperglycaemia and its associated complications by finger millet (*Eleusine coracana* L.) seed coat matter in streptozotocin-induced diabetic rats. *British Journal of Nutrition, 104*(12), 1787–1795. https://doi.org/10.1017/S0007114510002977

Shrestha, J., Subedi, S., Shrestha, S., Rai, S., Poudel, A., Adhikari, S., Baral, P., Pandey, M., Aryal, A., Kayastha, P., Limbu, A. K., Dhami, A., & Karanjit R. (2024). Crop biofortification of finger millet through agronomic approaches to enhance bioactive compounds for human health and nutrition. *Agronomy Journal of Nepal, 8*, 237–249. https://doi.org/10.3126/ajn.v8i1.70921

Siegel, R. L., Miller, K. D., Fuchs, H. E., & Jemal, A. (2021). Cancer statistics, 2021. *CA: A Cancer Journal for Clinicians, 71*(1), 7–33. https://doi.org/10.3322/caac.21654

Singh, N., Soni, S., Noor, U., & Gupta, E. (2024). Neutraceuticals profile and health benefits of finger millets. *Nutraceutical Potential of Millets, 1*, 1–17.

Singh, V., Lee, G., Son, H., Amani, S., Baunthiyal, M., & Shin, J. H. (2022). Anti-diabetic prospects of dietary bio-actives of millets and the significance of the gut microbiota: A case of finger millet. *Frontiers in Nutrition, 9*, 1056445. https://doi.org/10.3389/fnut.2022.1056445

Tanwar, E., & Kalpana, K. (2024). Can consumption of finger millet diet improve mental health status in athletes: A possible link with modulation of cortisol levels. *Physical Activity and Nutrition, 28*(4), 49–59. https://doi.org/10.20463/pan.2024.0032

Udeh, H. O., Duodu, K. G., & Jideani, A. I. O. (2017). Finger millet bioactive compounds, bioaccessibility, and potential health effects - a review. *Czech Journal of Food Sciences, 35*(1), 7–17. https://doi.org/10.17221/206/2016-CJFS

Vellaiyan, D., Arumugam, T., Shenbagavalli, S., Sangeetha, S., Anitha, R., & UmaMaheshwari, T. (2023). A review on nutritional properties and health benefits of finger millet. *International Journal of Plant & Soil Science, 35*, 753–761. https://doi.org/10.9734/ijpss/2023/v35i183342

Warhade, S., Atram, P., & Sawarkar, N. (2023). Finger millet (ragi): A nutritional boon for osteoporosis. *International Journal for Research in Applied Science & Engineering Technology, 11*(7), 562–564. https://doi.org/10.22214/ijraset.2023.54700

WHO. (2024). *Obesity and overweight*. https://www.who.int/news-room/fact-sheets/detail/obesity-and-overweight

Zhang, L., La, X., Tian, J., Li, H., Li, A., Liu, Y., Wu, C., & Li, Z. (2021). The phytochemical vitexin and syringic acid derived from foxtail fillet bran inhibit breast cancer cells proliferation via GRP78/SREBP-1/SCD1 signaling axis. *Journal of Functional Foods, 85*, 104620. https://doi.org/10.1016/j.jff.2021.104620

9

Finger Millet in Prebiotic and Probiotic Foods/Genetic Improvement

Palak Mahajan and Sohani Dadwal

9.1 Introduction

Despite continuous industrialization and urbanization, health challenges are one of the major concerns faced worldwide. The progress and development of the nation truly and wholly depend on the physical and mental health of its citizens. The people of developing and underdeveloped countries are still dealing with malnutrition, which is a root cause of various serious health problems. Malnutrition is one of the most serious concerns, as undernourished children are more vulnerable to the "malnutrition-infection cycle" due to their compromised immune systems (Siddiqui et al., 2021). This problem cannot only be coped by the availability of wholesome food; however, strong intestinal health is also required to increase the bioaccessibility and bioavailability of the required nutrients. This can be achieved by maintaining good health of the intestinal tract, which is considered the second brain of the human body. The intestinal flora helps in digestion and absorption of the nutrients from food and produces functional compounds necessary for life such as vitamins, short-chain fatty acids (SCFA), and amino acids (Vernocchi et al., 2020). However, if the flora is disturbed then this can affect human health and even cause problems such as diabetes, obesity, and colon cancer (Nie et al., 2018). This thus makes it necessary to regulate the important stakeholders of the intestinal tract, i.e., prebiotics and probiotics.

Probiotics are live microbes that benefit the human body by inhibiting the colonization of pathogenic bacteria, and helping to build the protective layer of intestinal mucosa resulting in improved immunity (Bodke & Jogdand, 2022). However, the growth and reproduction of the probiotics is strongly influenced by the prebiotics, which are food components, specifically polysaccharides and oligosaccharides that cannot be digested and absorbed by the human body and can be decomposed by the microflora

 DOI: 10.1201/9781003535119-9

present in the gastrointestinal tract (GIT), generally by *Lactobacilli* and *Bifidobacteria*. Upon utilization by these colonic microflorae, the prebiotics form SCFA, which are dispersed in the colon and afterward absorbed into the bloodstream. The merits of prebiotics for health include refinements in cardio-metabolism, lowered blood lipid levels, magnified insulin sensitivity, improved bone health, brain function and cognitive abilities, and enhanced neurovegetative activity and antioxidant functions (Bevilacqua et al., 2024). These health benefits of prebiotics are connected with the interlinkage between the microbiome and the host, influencing microbiota balance, the gut-brain axis, and communication between the microbiota and the immune system, as well as the utility of the intestinal barrier (Nobre et al., 2022). To make the proper use of prebiotics, probiotics primarily consisting of Gram-positive bacteria, hold species from the genera *Lactobacillus* and *Bifidobacterium* are responsible. Thus, probiotics also prevent many clinical conditions, advance the intestinal microenvironment, facilitate immune regulation, alleviate physiological stress, inhibit pathogen expansion, and enhance the barrier function of the intestinal epithelium, among others (Ma et al., 2022). Probiotic products are provided in various forms globally, such as capsules, powders, food items, and infant formulas as nutraceuticals. Prebiotics have been embodied in many different food products, including fermented milk and yogurts, breakfast cereals, health and sports beverages, energy bars, sugar-free confections, chewing gum, baked items, and meat products (Precup et al., 2022). Despite the various products available, cereal-based foods form a larger portion of the food products consumed worldwide. Fermentation of the cereals by lactic acid bacteria not only reduces the antinutritional compounds such as phytates and tannins but also increases protein availability. Millets, pseudo-cereal grains, form one of the most environment-friendly grains and have remained the staple food for the lower socioeconomic people. The millet fibers such as resistant starch and beta-glucans, generate a favorable environment for the hike of the probiotics. Consolidating millets into probiotic-rich products, such as fermented dairy or plant-based blends, promotes the nutritional value of these foods while contributing the genetic prebiotic properties of millets to enlarge the effectiveness of probiotics. Among the various major and minor millet, finger millet (*Eleusine coracana* L.), commonly known as ragi, holds an eminent place worldwide regarding both its production and utilization. This grain transcends rice and wheat in terms of its mineral, fiber, and micronutrient content (Karki et al., 2020). Thus, millets are acquiring perception as relevant substrates for prebiotics and probiotics due to their excellent nutritional and biochemical characteristics. Thus, this chapter will focus on the various types of prebiotics and probiotics, their role, and their mechanism of action. The role of finger millet as a prebiotic and probiotic food is also discussed in detail.

9.2 Types of Prebiotics and Probiotics

Previously prebiotics were considered as a carbohydrate component mainly xylooligosaccharides, galacto-oligosaccharides, inulin, lactulose, and fructooligosaccharides (de Paulo Farias et al., 2019). However, recent studies have now included non-carbohydrates such as polyphenols and polypeptides as a part of the prebiotic group. To classify a compound as a prebiotic it must (i) resist the acidic environment of the stomach, (ii) not be degraded by the digestive enzymes, (iii) not be absorbed in the intestinal tract, and (iv) fermented by the probiotics or intestinal microflora (Davani-Davari et al., 2019). The various emerging prebiotic source, components, and their functions are listed in Table 9.1. Galacto-oligosaccharides are one of the most common and widely used prebiotics with many health-promoting benefits such as regulating the balance of the human colon microflora and stimulating the proliferation of *Bifidobacterium* in the intestine. In addition, the inulin-type fructans are another common prebiotic that promotes the growth of probiotic flora. Apart from its function as a prebiotic, inulin offers additional benefits as it can replace fats and carbohydrates providing the desired food flavor, it also promotes mineral absorption, prevents gastrointestinal disease, and boosts the immune system. The effect of the prebiotic on human health is shown in Fig. 9.1.

Probiotics have a wide range of species, mainly *Lactobacilli* and *Bifidobacteria* in addition to others. Many research articles have focused mainly on the LAB group. *Lactobacillus*, an important probiotic, promotes the synthesis and absorption of essential vitamins, amino acids, and minerals, inhibits the growth of harmful microorganisms, and improves intestinal health. SCFA promotes the growth and reproduction of *Lactobacillus* and inhibits *Escherichia coli*, thus supporting the functioning of the colon (Markowiak-Kopeć & Śliżewska, 2020). *Bifidobacterium*, a genus of gram-positive bacteria can adapt to the anaerobic environment of the intestine and thus can easily reproduce and metabolize in the middle and ends of the small and large intestine. It restricts the growth of the pathogenic microbes and promotes the balance of normal intestinal bacterial flora. In addition, *Bifidobacterium bifidum* showed anti-tumor properties (Abdolalipour et al., 2020). Besides these, *Enterococcus*, *Saccharomyces cerevisiae*, and *Saccharomyces boulardii* are also often used in the food industry. *Enterococcus* is highly resistant to a wide range of pH and temperature and is also used as an antimicrobial agent (Hanchi et al., 2018). *Saccharomyces cerevisiae*, a non-pathogenic probiotic is now commercially used for the production of probiotic foods (Souza et al., 2021). *Saccharomyces boulardii* possesses probiotic properties, helps to maintain the balance of normal intestinal microflora, and is commonly used to treat digestive disorders (Pais et al., 2020).

TABLE 9.1

Utilization of Finger Millet as Prebiotic and Probiotic Foods

Product Name	Raw Material	Processing Parameters	Key Characteristics	References
Fermented milk	• Finger millet • *Streptococcus thermophilus* MTCC 5460 • Probiotic strain *Lactobacillus helveticus* MTCC 5463 • Carboxymethylcellulose	• 37°C ± 1°C for 6 hours • Stored at (5 ± 2)°C	The product may have potential as a functional food for diabetes treatment and prevention. • It also successfully reduced polyphagia and polydipsia.	Chaudhary and Mudgal (2020)
Probiotic beverages	• *Lactobacillus plantarum* MF405176.1 • *Lactobacillus curieae* MF405178.1 • Finger millet • Banana flour substrates	• 37°C ± 1°C for 24 hours	• The developed probiotic beverage demonstrated improved physical and chemical characteristics, increased acceptance, and better shelf life.	Divisekera et al. (2021)
Vegan probiotic beverage	• White finger millet (WFM) • Pineapple crown powder (PCP) • *Lactobacillus rhamnosus* LGG	37°C ± 1°C for 6 hours	• Increase in proximate composition such as ash, fat, and crude fiber	Byresh et al. (2023)
Probiotic bacteria isolates	• MRS (de Man Rogosa and Sharpe) • 0.1% sodium hypochlorite and 70% ethanol	Temp: 17°C, 27°C, and 40°C pH: 4.5, 6.5, and 8.5	Probiotic properties, antibacterial and antifungal activity, acid and bile salt tolerance, antagonistic activity against pathogens, and different antibiotic sensitivity profiles	Gouthami et al. (2024)
Probiotic beverage	• WFM flour • Microbial inoculum • Sugar	• Millet flour: 10–14% w/v, • microbial inoculum: 1–5% v/v, • sugar: 5–15% w/v	Increase in the flour and inoculum content showed significant effect on the pH value and microbial load. Overall, the product can be used as an alternative for consumers with strict vegan and lactose-intolerant diet.	Navyashree et al. (2023)
Probiotic beverage	• Cooked finger millet • *L. casei*	• Inoculation temperature: 37°C • Time: 2, 4, and 6 hours	Lowered the reducing and non-reducing sugars contents	Fasreen et al. (2017)
Fermented beverage	• Finger millet • Co-culture: *Weissella confusa* 2LABPT05 and *Lactiplantibacillus plantarum* 299v	• 37°C overnight	High content of fiber, dextran, protein digestibility and limiting amino acid and low content of sugar and fat	Vila-Real et al. (2022)

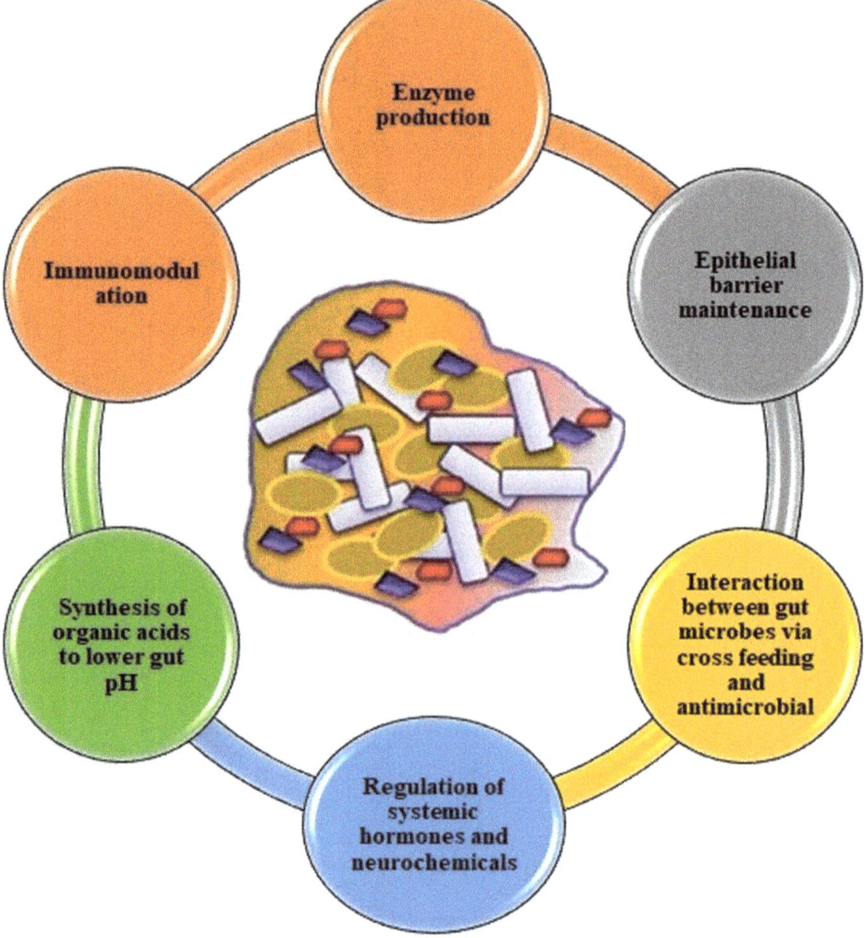

FIGURE 9.1
Role of prebiotics on human health.

9.3 Mechanism of Action of Prebiotics and Probiotics

Generally prebiotic food escapes digestion due to a lack of hydrolysis enzymes in the human body and remains in the small intestine. These prebiotics are then transported to the large intestine where they are degraded by the intestinal flora and fermented to produce certain secondary metabolites absorbed by the intestinal epithelium and then transported to the liver where they possess health benefits such as regulation of immunity, resistance to pathogens, enhanced mineral absorption, and reduction in lipid profile (You et al., 2022).

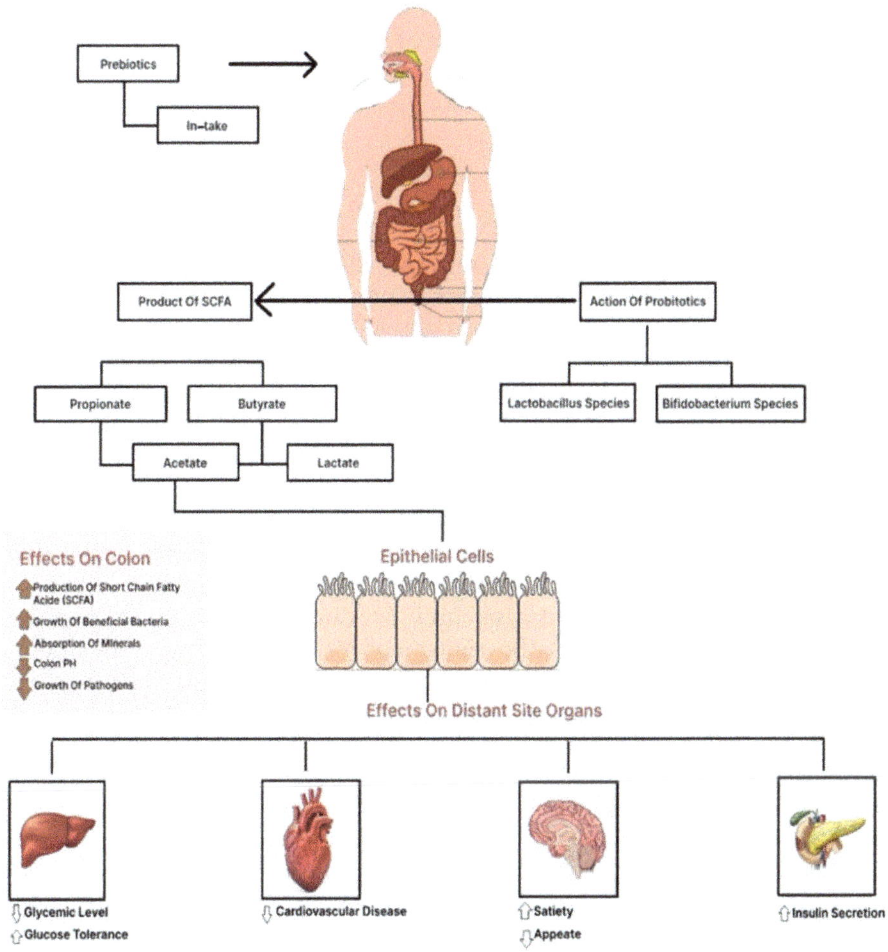

FIGURE 9.2
Mechanism of prebiotic and probiotic action and their effect on human health.

Source: Adapted and modified from Abdi and Joye (2021).

The beneficial bacteria then metabolize the SCFAs formed in the intestine to beneficial compounds such as acetate, propionates, and butyrates, responsible for maintaining health (Fig. 9.2). Consumption of these prebiotics promotes the growth of the beneficial flora, which not only supports gastrointestinal health but also the healthy functioning of other body parts as cardiovascular, immune, and central nervous systems. The mechanism of action of the probiotic is shown in Fig. 9.2. The mucus layer and intestinal epithelium act as a physical barrier against the intestinal harmful microorganism, facilitate the food passage, and restrict the adhesiveness of the pathogenic microbes onto the lamina propria. Probiotics enhance the barrier properties, restore intestinal permeability, and enhance mucus production and integrity.

9.4 Finger Millets in the Prebiotic and Probiotic Foods

Finger millet is a good source of resistant starch and dietary fiber which make it a good substrate for the probiotic strains. The utilization of the finger millet in some of the products is shown in Table 9.1. The low glycemic index and good prebiotic activity promote gut health, digestion, and bioavailability of nutrients (Singh et al., 2022). Yogurt is an effective vehicle for incorporating sources of dietary fiber into meals. Finger millet, a rich source of dietary fiber, was added to the yogurt. *Lactobacillus acidophilus* and *Bifidobacterium animalis* as probiotic cultures and *Streptococcus thermophilus* and *Lactobacillus delbruckii* as starter cultures were used. Along with maintaining the sensory profile and physicochemical properties, the addition of finger millet flour also gave negative results to the coliforms (Disanth et al., 2022).

However, non-dairy beverages and products have gained popularity across the globe due to the rising concerns regarding milk allergy, lactose intolerance, high calories, and hypercholesterolemia. The most frequently used strains are lactic acid bacteria, *Saccharomyces boulardii,* and *Bifidobacteria,* and the other potential strains are *Streptococcus, Bacillus,* and *Enterococcus.* A study evaluated the prebiotic potential of the xylooligosaccharides extracted from the seed coat of the finger millet (Palaniappan et al., 2017). The bacterial growth of *Lactobacillus plantarum, Lactobacillusbrevis, Lactobacillus. sakei,* and *Lactobacillus acidophilus* on the extracted xylooligosaccharide was found comparable to the dextrose which confirmed its prebiotic potential. Also, the utilization of xylooligosaccharides was different among the strains with the highest score for *Lactobacillusplantarum,* which was attributed to their homofermentative property. A probiotic functional beverage was developed from the white finger millet (WFM), pineapple crown powder (PCP), and *Lactobacillus rhamnosus* GG NCDC 347 (LGG) as probiotic microbes (Byresh et al., 2023). The optimized sample showed an overall increase in the nutritional composition, total phenolics (10.02 mg GAE/g), flavonoid content (8.24 mg QE/g), antioxidant, physiochemical, and functional characteristics which suggest the potential of the sample for medicinal uses. Similarly, a vegan WFM probiotic beverage (WFMPB) using WFM flour (10–14% w/v), microbial inoculum (1–5% v/v), and sugar (5–15% w/v) was developed. The study also suggested the use of finger millet as a substitute for consumers who are lactose-sensitive or strictly vegan (Navyashree et al., 2023). In another study, to enhance the nutritional properties and to utilize the by-product, pineapple core powder was added to the WFM flour, and *Lactocaseibacillus rhamnosus* was used to develop plant-based probiotic beverages. The addition of pineapple core powder improved the nutritional value of millet drinks, which is a beneficial outcome (Malini et al., 2024).

In this regard, the two widely employed techniques for improving the prebiotic and probiotic potential of plant-based sources, mainly cereal and millet are germination and prebiotic fermentation. These techniques result in structural modification, reduced antinutritional factors, and increased digestibility, bioactivity, and nutritional profile as shown in Fig. 9.3.

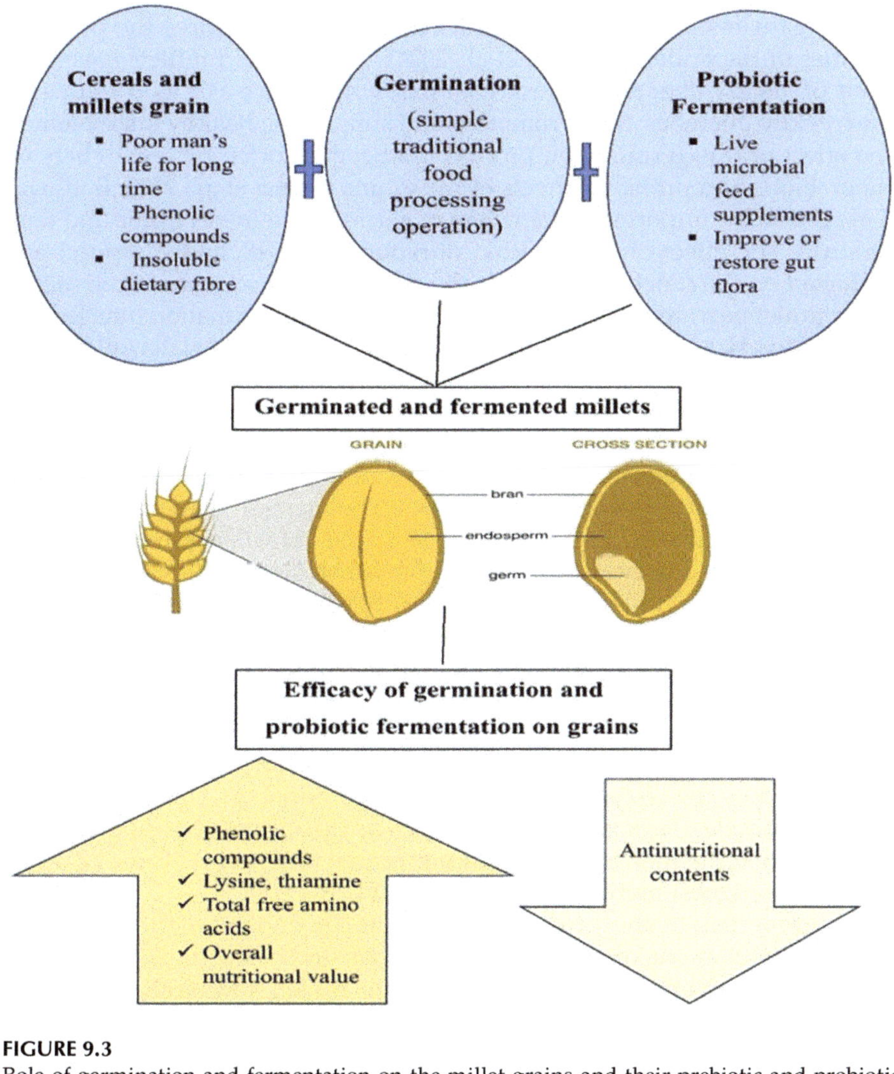

FIGURE 9.3
Role of germination and fermentation on the millet grains and their prebiotic and probiotic activities.

9.4.1 Effect of Germination

Germination, also interchangeably known as sprouting, is a complex phenomenon that involves the activation of various enzymes and metabolic pathways within the grain seed, which degrades the storage reserves such as starch and proteins to provide energy and nutritional compounds for plant growth (Marti et al., 2020). The process is also known to stimulate the synthesis of health-promoting secondary metabolites such as phenolics. The process has been used since ancient times to increase nutritional digestibility and bioavailability by reducing antinutritional compounds such as phytates, tannins, and oxalates. Moreover, germination enhances the proliferation of microorganisms, mainly lactic acid bacteria, which enhances the prebiotic qualities of the millets (Kumari et al., 2024). The sprouted millets have high levels of dietary fiber and oligosaccharides, which serve as a food source for gut-friendly microbes, thus promoting and supporting healthy microbiomes. The effect of this germination process has been reported by researchers on the prebiotic and probiotic effects of the grains (Sneha et al., 2023; Budhwar et al., 2020). The important parameters of germination involve time and temperature. The effect of germination duration (0, 24, 48, and 72 hours) was evaluated on the functional, physicochemical, and sensory characteristics of finger millet porridge and flour. The increase in the germination time led to a decrease in pH, an increase in titrable acidity, and improved water/oil absorption capacities and solubility. Overall, the germination led to improved physicochemical, functional, and sensory characteristics of finger millet flour and porridge (Nefale & Mashau, 2018). Similar to this, a study determined the effect of germination duration (24, 48, and 72 hours) on the physicochemical characteristics of finger millet varieties. The increase in the L value, protein content, titrable acidity, and decrease in the pH value, ash content, and fat content were observed on germination which significantly indicated its effect on the finger millet flour (Yenasew & Urga, 2022). The other important factor is temperature. Sneha et al. (2023) evaluated the effect of germination temperatures (22°C, 26°C, and 30°C) for various time durations as 24, 48, and 72 hours on the prebiotic effect of finger millet. The prebiotic effect of the germinated grain was higher compared to the non-germinated one because of the presence of the components suitable for optimal growth of bacteria synthesized by the enzymatic process upon germination. Upon inoculation with *Lactobacillus plantarum* and *Lactobacillus acidophilus*, the finger millet germinated at the highest temperature for more time, i.e., 30°C for 72 hours showed the highest prebiotic effect of 9.23 CFU/g.

The properties of the probiotic drinks can be enhanced further by the use of more than one dietary fiber or resistant starch source. In this regard, germinated and ungerminated barley, finger millet, and moth bean seeds were used to create probiotic drinks in the 2.5:1.5:1 for barley: finger millet: moth bean. These drink concoctions (0, 2, 4, 6, and 8 g) were added to distilled water and soy, almond, and coconut milk. The probiotic bacterium

Lactobacillus acidophilus was added to drink combinations. Variations in the concentration of drink mixture in three distinct milk and distilled water were accompanied by increases in the pH and acidity, bacterial count, 2,2-Diphenyl-1-Picrylhydrazyl (DPPH) assay, and polyphenol content (Chavan et al., 2018). Further, to increase the functionality of the germinated finger millet, sporicidal treatments using various dry and wet methods were tested to destroy the spores, which would otherwise compete with starter culture during the fermentation process and prevent the production of probiotic millet food (Jayashree et al., 2024). After autoclaving finger millet for 30 minutes at 121°C, the number of aerobic spores decreased from 3.86 to 1.00 log_{10} CFU/g. This further suggested autoclaving as a sporicidal treatment for probiotic millet meal. Overall, the germination process showed the prebiotic and probiotic ability of the finger millet in addition to the other benefits, such as improved digestibility and nutrient bioavailability.

9.4.2 Effect of Fermentation

Fermentation further amplifies the prebiotic and probiotic potential of the millets. During the process, naturally occurring microorganism breaks down the complex carbohydrates into simple compounds. This metabolic activity not only improves digestibility by reducing and degrading the antinutritional compounds but also enhances the production of bioactive compounds. This promotes the balance of good microbes which enhances the gut barrier function, improves nutrient absorption, and reduces inflammation. Various studies reported the benefits and efficiency of fermented finger millet and its products for their prebiotic and probiotic activity (Kumar et al., 2022; Di Stefano et al., 2017). For instance, a study reported a gradual increase in the probiotic cell count upon increasing the fermentation time for the finger millet and banana flour based dairy free probiotic beverage using Lactic acid bacterias, *Lactobacillus plantarum* and *Lactobacillus curieae* (Divisekera et al., 2021). Among all the beverages, the finger millet-based probiotic beverage containing *L. plantarum* showed better physical, and chemical attributes and higher acceptability. The study thus concluded that the fermented finger millet is compatible as suitable prebiotic substrates for the manufacture of probiotic food, as proven by its ability to sustain the viability of probiotic starter cultures up to 10^9 CFU/ml even at the end of the fifth week of shelf life. In addition to these, the lactic acid bacteria isolated from the fermented finger millet also showed potential anticancerous activity. *Lactobacillus plantarum, Lactococcus lactis, Limosilactobacillus fermentum, Enterococcus faecium,* and *Pediococcus. acidilactici* showed superior probiotic attributes like pH tolerance, phenol tolerance, osmo tolerance, bile tolerance, and temperature tolerance (Divisekera et al., 2019). Similarly, bioactive isolates (*Lactiplantibacillus plantarum* MYSRD57, *Lacticaseibacillus paracasei* MYSRD68, *Lactiplantibacillus plantarum* MYSDS3, *Lactiplantibacillus plantarum* MYSDS4, and *Lactiplantibacillus plantarum* MYSDVN3) from finger millet batter, a traditional fermented dish showed significant probiotic traits, including tolerance toward acidic pH, 0.3% bile, NaCl, phenols, and temperature conditions, resistance to antibiotics, and anti-hemolytic action (Shivakumar et al., 2023). The fermented finger millet milk also showed a reduction in the antinutritional factors and pH, while phenolic content and antioxidant activity increased (Yadav et al., 2023).

The functionality of the finger millet can be improved with the addition of the food material with the symbiotic relation. In this regard, the co-culture of exopolysaccharide-producing *Weissella confusa* and prebiotic strain *Lactiplantibacillusplantarum* showed high fermentative performance on the nutritionally rich finger millet (Vila-Real et al., 2022). This symbiotic consortium led to the development of a novel functional finger millet-based yogurt-like beverage with high viable cell numbers of probiotic microorganisms. The product showed high fiber and dextran content, low sugar and fat content, increased protein digestibility, and enhanced amino acid profile. This makes the developed product a healthy food product to snack on among the different aged population groups, and those suffering from lactose or gluten intolerance. Moreover, the finger millet-based probiotic beverage developed by incubated cooked finger millet inoculated with *Lactobacilluscasei* at 37°C for 2, 4, and 6 hours can also reduce the content of reducing and non-reducing sugars (Fasreen et al., 2017). Not only this, but the traditional fermented multi-grain millet dosa batter also showed probiotic qualities. Six isolates from the batter, i.e., *Lactobacillus pentosus, Lacticaseibacillusparacasei, Lactiplantibacillus plantarum*, and *Limosilactobacillus fermentum* were able to withstand for up to five hours both 0.3% bile salt (5%–80% viability) and acidic environments (45%–80% viability). They resisted phenol (53.3%–83.6% viability) and NaCl (38.2%–49.5% viability), decreased cholesterol levels, strong adhesion to epithelial and buccal cells, and robust survival in gastric juice simulations. In addition, their antibacterial action was noted against several pathogens, such as *Escherichiacoli* and *Salmonella typhi* (Divyashree et al., 2024).

Ultrasound treatment can enhance the functional activities of the probiotic and prebiotics. WFM was treated with ultrasound both before and after the inoculation. After the fermentation, the ultrasound-treated samples showed a significant increase in the total phenol content, total flavonoid content, antioxidant activity, and reduced sugar concentration (Meena et al., 2023). the treatment also causes structural modifications to millet protein and other components, which have an impact on the techno-functional behavior of MPB, including viscosity and water-holding capacity. These results demonstrate that the US treatment has a significant impact on the probiotic microorganisms' fermentation kinetics and that it can be used to modify the physicochemical alterations in probiotic drinks (Meena et al., 2024).

Overall, finger millet presents a promising potential to be used as a functional substrate for the prebiotic and probiotic foods, owing to its high levels of the dietary fiber, nutritional profile, and bioactive compounds. As the consumer demand grows for the functional food, the addition of the finger millet into the prebiotic and probiotic formulations could play an important role in the promotion of the healthier gut microbiomes. With the further advancement in this arena, the finger millet could possess the potential to be a cornerstone ingredient in the functional food industry.

9.5 Genetic Improvement of Finger Millet

Despite its nutritional value, finger millet remains a minor crop due to various agronomic constraints and limited processing adaptability. Finger millet is vulnerable to the climacteric risks despite being considered as a climate resilient crop which can drive food cultivation and production for the developing world. Climate change is happening quickly on a global scale which has greatly impacted agriculture production (Wambi et al., 2021). For example, several molecular changes are induced under the high temperature conditions which can adversely impact the major biological processes such as transpiration, respiration and photosynthesis. The rise in the global temperature can threaten the food security by influencing the food productivity. Finger millet cultivars are grown widely in the regions with the highest temperatures of 36°C, while a further rise can adversely impact its yield. Similar to other crops, the biological processes in finger millet are also temperature-sensitive (Yogeesh et al., 2016). Additionally, the rise in CO_2 indirectly impacts the performance of the plant by affecting the plant's metabolism. Moreover, the evidence showed that the amount of land suitable for crop production including finger millet will decline worldwide because of abiotic stresses such as drought, salinity, temperature fluctuations, and nutrient deficiency. To protect the crops, form their productivity decline, biotechnological tools such as genomic selection, hybridization, cross breeding, and genetic modifications are required. Molecular markers such as simple sequence repeats, random amplification of the polymeric DNA, expressed sequenced tags, and inter simple sequence repeats can be used to access the genetic variability in the millet crop (Sood et al., 2016). These genetic improvement strategies are key to enhancing yield, resistance to biotic (pest, pathogens) and abiotic (temperature, salinity, solar radiation) stresses, and food processing characteristics (Gupta et al., 2017).

Indian nation gene bank holds the largest collection of 10,507 accessions including exotic and rest indigenous lines and wild species. The genetic diversity specifically the morphological, present in the finger millet is immense. For instance, finger millet with a range of seed colors can be produced. Landraces with various characteristics such as drought tolerance, maturity time, and bird tolerance are valued by the farmers impacting the productivity via multiple agroeconomic zones (Tsehaye et al., 2006). Thus, genetic diversity serves tremendous opportunities to manipulate the required traits.

The sustainable production of the finger millet requires a transition to the agricultural production systems generating crops that are resilient to climate change. The use of the finger millet genetic and genomic resources is one of the pathways to the climate-smart approach that could provide sustainable and cost-effective options which play a crucial role in increasing long-term, crop productivity. The cultivated finger millet (*Eleusine coracana* subsp. *Coracana*; 2n = 4x = 36) is a large-grained mutant that exhibits a wider distribution (Wambi et al., 2021). This presents the opportunity for immense and evolutionary adaptive genetic variations. The studies have identified various beneficial variations leading to the availability of the germplasm which can mature early, yield more, and is blast resistant. A total of 58 simple sequence repeats markers related to the blast resistance have been identified in the finger millet (Babu et al., 2014). Moreover, some drought and salinity-tolerant accessions have been reported that can be used in the hybridization programs to strengthen the finger millet resilience toward climatic stresses. Salinity responsive candidate genes, for instance, mineral exchangers, signal transducers, transporters and stress tolerants have been identified through RNA sequencing in the finger millet. Under abiotic stress, *EcbZIP17* showed increased stress and growth tolerance (Ramakrishna et al., 2018). From the last decade, research has been done to identify and develop the genotype with the tolerance toward climatic stresses.

An accurate assessment of the genetic diversity of the finger millet germplasm and its population structuring is crucial for the genetic resource management, mapping of traits and crop improvement. Further it has been reported that the extreme drought conditions can adversely impact the uptake of the nutrients, for example, nitrogen and phosphorus in the plants (Bista et al., 2018). However, the nutrient signaling breeding in the finger millet has not yet been exploited due to the lack of molecular information available related to the nutrient uptake. For the development of the low nutrient tolerant finger millet cultivar, a depth understanding and knowledge of the nutrient deficiency is required. Thus, a complete genome aided research is expected to help in the precise estimation of the tolerance of nutrient starvation. For instance, numerous candidate genes such as *EcGS*, *EcDof1*, *EcHNRT2*, and *EcLNRT1* have been identified which can be used to develop nitrogen use efficient finger millet crops (Gupta et al., 2013). Similarly, candidate gene *EcDehydrin7* can be used for the development of the drought and heat tolerant finger millet variety (Singh et al., 2014). This will eventually aid in selecting of the cultivars for the regions where nutrient deficiency is noticeable. This will thus help in the enhanced productivity with decreased dependency on the expensive fertilizers and thus ultimately reducing the greenhouse gases emission. Further, first transgenic finger millet conferred resistance to the response of the *Pyricularia grisea* (Latha et al., 2005). Similarly, another transgenic finger millet developed via Agrobacterium-mediated transformation exhibited resistance toward leaf blast disease (Ignacimuthu & Ceasar, 2012). Some transgenic finger millet showed salt and drought tolerance.

In addition to the improvement in the agronomic traits such as high yield, pest resistance, and climate stress tolerance, the genetic improvement of the finger millet can play a crucial role in enhancing the food applications. With the increase in the rise in the demand of the gluten-free food products, the finger millet can be a viable alternative to the wheat in the baking and other food products (Selladurai et al., 2023). Via the genetic improvement, finger millet varieties with the better milling and baking properties can be developed to produce gluten-free flour for wider food applications. Furthermore, nutrient profile of the finger millet can be enhanced by the genetic improvements (Sood et al., 2019). Specific nutritional deficiencies such as iron, zinc or other essential amino acid in the population can be addressed by the breeding of the finger millet. Similarly, genetic improvements for the processing traits such as reduction in the antinutritional factors can also improve the digestibility and nutrient absorption of these products.

The wide use of the finger millet genomics and genetic resources via climate smart breeding is one of the pathways which can significantly contribute in this regard. Proteomics is a popular omics approach that is widely used to understand the climate resilience and nutrition related traits. The approach helps in the characterization of the molecular components of heterosis which would help in the improvement of the crops. Calcineurin-B and calreticulin proteins have been identified in the seeds of the finger millet which are involved in retaining the high contents of calcium (Singh et al., 2015). Once such proteins can be fully characterized then they can be used for the food products rich in calcium (Sood et al., 2016). Transcriptomics studies have also provided the basis understanding of the nutritional variations, and tolerance toward salinity and drought conditions. Integration of these data can ease the identification of the suitable candidate genes that can be manipulated for the development of the improved millet crops or crops in general. The sequence information is readily available for the carious crop species but still minor crops still lag behind and finger millet is one of such orphan crops. However, with the advancement in the technologies, it has become easy to generate the required data for the computational gene and molecular market discovery. Linkage of these markers with the gene traits can be employed to associate the genotype with the expressed phenotype. This would benefit the population as well as farmers in combating the food security as well as producing a crop with advanced nutritional benefits.

9.6 Conclusion

The probiotic and prebiotic foods available in the market are mostly dairy-based. Demand for unusual and unique flavors, an increased population of vegan consumers, and allergies and intolerance toward lactose have led to few attempts to generate this probiotic and prebiotic food utilizing substrates such as cereals and millets. The widespread availability of these substrates and their high nutritional profile have drawn attention to their usage for these roles. Millets are storehouses for dietary fiber and resistant starch which makes them suitable substrates for probiotic activity. The other constituents involved are polyphenols, minerals, and vitamins which offer additional health benefits. The studies have reported the development of various probiotic beverages and foods that could help developing countries overcome their economic challenges as well as the high cholesterol and lactose intolerance of dairy products in rich nations. However, future research is still required to investigate the effect of novel techniques on probiotic and prebiotic food based on finger millet or millet in general. This could further open new paths for the millets to enter in the probiotic and prebiotic arena.

Further the genetic improvement of the finger millet is a powerful strategy to enhance the role and application of the finger millet in the global food markets, especially the regions dealing with the food insecurity, malnutrition, and climate related issues. Finger millet varieties possessing the benefits of higher yields, better nutritional profiles and improved food processing qualities can be developed via biotechnological innovations and hybridizations. As the demand of the healthier, sustainable, and more varied food product is growing, the genetic improvement of the finger millet offers a promising pathway to not only increase its productivity but also unlocks the pathways to explore its full potential in the food industry.

References

Abdi, R., & Joye, I. J. (2021). Prebiotic potential of cereal components. Foods, 10(10), 2338. https://doi.org/10.3390/foods10102338

Abdolalipour, E., Mahooti, M., Salehzadeh, A., Torabi, A., Mohebbi, S. R., Gorji, A., & Ghaemi, A. (2020). Evaluation of the antitumor immune responses of probiotic bifidobacterium bifidum in human papillomavirus-induced tumor model. Microbial Pathogenesis, 145, 104207. https://doi.org/10.1016/j.micpath.2020.104207

Babu, B. K., Agrawal, P. K., Pandey, D., Sood, S., Chandrashekara, C., & Kumar, A. (2014). Molecular analysis of world collection of finger millet accessions for blast disease resistance using functional SSR markers. SABRAO Journal of Breeding and Genetics, 46(2), 202–216.

Bevilacqua, A., Campaniello, D., Speranza, B., Racioppo, A., Sinigaglia, M., & Corbo, M. R. (2024). An update on prebiotics and on their health effects. Foods, 13(3), 446. https://doi.org/10.3390/foods13030446

Bista, D. R., Heckathorn, S. A., Jayawardena, D. M., Mishra, S., & Boldt, J. K. (2018). Effects of drought on nutrient uptake and the levels of nutrient-uptake proteins in roots of drought-sensitive and-tolerant grasses. Plants, 7(2), 28. https://doi.org/10.3390/plants7020028

Bodke, H., & Jogdand, S. (2022). Role of probiotics in human health. Cureus, 14(11). https://doi.org/10.7759/cureus.31313

Budhwar, S., Sethi, K., & Chakraborty, M. (2020). Efficacy of germination and probiotic fermentation on underutilized cereal and millet grains. Food Production, Processing and Nutrition, 2, 1–17. https://doi.org/10.1186/s43014-020-00026-w

Byresh, T. S., Malini, B., Kavitha, L., Sunil, C. K., Chidanand, D. V., & Venkatachalapathy, N. (2023). Study of prebiotic effect of pineapple crown powder on development of white finger millet vegan probiotic beverage. Food and Humanity, 1, 742–752. https://doi.org/10.1016/j.foohum.2023.07.021

Chaudhary, J. K., & Mudgal, S. (2020). Antidiabetic and hypolipidaemic action of finger millet (*Eleusine coracana*)-enriched probiotic fermented milk: An in vivo rat study. Food Technology and Biotechnology, 58(2), 192. https://doi.org/10.17113/ftb.58.02.20.6308

Chavan, M., Gat, Y., Harmalkar, M., & Waghmare, R. (2018). Development of non-dairy fermented probiotic drink based on germinated and ungerminated cereals and legume. LWT, 91, 339–344. https://doi.org/10.1016/j.lwt.2018.01.070

Davani-Davari, D., Negahdaripour, M., Karimzadeh, I., Seifan, M., Mohkam, M., Masoumi, S. J., Berenjian, A., & Ghasemi, Y. (2019). Prebiotics: Definition, types, sources, mechanisms, and clinical applications. Foods, 8(3), 92. https://doi.org/10.3390/foods8030092

de Paulo Farias, D., de Araújo, F. F., Neri-Numa, I. A., & Pastore, G. M. (2019). Prebiotics: Trends in food, health and technological applications. Trends in Food Science & Technology, 93, 23–35. https://doi.org/10.1016/j.tifs.2019.09.004

Disanth, E., Anand Kumar, S., Susruthan, V., & Piratheepan, S. (2022). Effect of incorporation of finger millet (*Eleusine coracana*) on proximate composition, sensory and microbiological properties and shelf life of drinking yoghurt over refrigerated storage. University of Sabaragamuwa, Sri Lanka.

Di Stefano, E., White, J., Seney, S., Hekmat, S., McDowell, T., Sumarah, M., & Reid, G. (2017). A novel millet-based probiotic fermented food for the developing world. Nutrients, 9(5), 529. https://doi.org/10.3390/nu9050529

Divisekera, D. M. W. D., Samarasekera, J. K. R. R., Hettiarachchi, C., Gooneratne, J., Choudhary, M. I., Gopalakrishnan, S., & Wahab, A. T. (2019). Lactic acid bacteria isolated from fermented flour of finger millet, its probiotic attributes and bioactive properties. Annals of Microbiology, 69, 79–92. https://doi.org/10.1007/s13213-018-1399-y

Divisekera, D. M. W. D., Samarasekera, J. K. R. R., Hettiarachchi, C., Gooneratne, J., Gopalakrishnan, S., & Mazumdar, S. D. (2021). Development of probiotic beverages using finger millet [*Eleusine coracana* (L.) Gaertn.] and banana [*Musa* spp.] as prebiotic substrates. Asian Plant Research Journal, *8*(4), 123–131. https://doi.org/10.9734/APRJ/2021/v8i430193

Divyashree, S., Ramu, R., & Sreenivasa, M. Y. (2024). Evaluation of new candidate probiotic lactobacillus strains isolated from a traditional fermented food-multigrain-millet dosa batter. Food Bioscience, 57, 103450. https://doi.org/10.1016/j.fbio.2023.103450

Fasreen, M. M. F., Perera, O. D. A. N., & Weerahewa, H. L. D. (2017). Development of finger millet based probiotic beverage using *Lactobacillus casei*431. OUSL Journal, 12(1), 128–138. http://doi.org/10.4038/ouslj.v12i1.7384

Gouthami, B., Ramalakshmi, A., Balakrishnan, M., Karthikeyan, S., Muniraj, I., & Packia lakshmi, J. S. (2024). Functional and molecular characterization of millet associated probiotic bacteria. BMC Microbiology, 24(1), 485. https://doi.org/10.1186/s12866-024-03606-9

Gupta, A. K., Gaur, V. S., Gupta, S., & Kumar, A. (2013). Nitrate signals determine the sensing of nitrogen through differential expression of genes involved in nitrogen uptake and assimilation in finger millet. Functional & Integrative Genomics, 13, 179–190. https://doi.org/10.1007/s10142-013-0311-x

Gupta, S. M., Arora, S., Mirza, N., Pande, A., Lata, C., Puranik, S., Kumar, J., & Kumar, A. (2017). Finger millet: A "certain" crop for an "uncertain" future and a solution to food insecurity and hidden hunger under stressful environments. Frontiers in Plant Science, 8, 643. https://doi.org/10.3389/fpls.2017.00643

Hanchi, H., Mottawea, W., Sebei, K., & Hammami, R. (2018). The genus *Enterococcus*: between probiotic potential and safety concerns—An update. Frontiers in Microbiology, 9, 1791. https://doi.org/10.3389/fmicb.2018.01791

Ignacimuthu, S., & Ceasar, S. A. (2012). Development of transgenic finger millet (*Eleusine coracana* (L.) Gaertn.) resistant to leaf blast disease. Journal of Biosciences, 37, 135–147. https://doi.org/10.1007/s12038-011-9178-y

Jayashree, H., Malashree, L., Devaraju, R., Kumar, A., Pushpa, B. P., & Ramachandra, B. (2024). Sporicidal treatments to produce germinated finger millet. Journal of Scientific Research and Reports, 30(6), 586–592. https://doi.org/10.9734/jsrr/2024/v30i62076

Karki, A., Chandra, A., Joshi, S., Rawat, P., & Sharma, S. (2020). An overview of finger millet (*Eleusine coracana* L.). Journal of Pharmacognosy and Phytochemistry, 9(4), 866–869.

Kumar, D., Lal, M. K., Dutt, S., Raigond, P., Changan, S. S., Tiwari, R. K., & Singh, B. (2022). Functional fermented probiotics, prebiotics, and synbiotics from non-dairy products: A perspective from nutraceutical. Molecular Nutrition & Food Research, 66(14), 2101059. https://doi.org/10.1002/mnfr.202101059

Kumari, K., Kashyap, P., & Chakrabarti, P. (2024). Germination and probiotic fermentation: A way to enhance nutritional and biochemical properties of cereals and millets. Food Science and Biotechnology, 33(3), 505–518. https://doi.org/10.1007/s10068-023-01401-2

Latha, A. M., Rao, K. V., & Reddy, V. D. (2005). Production of transgenic plants resistant to leaf blast disease in finger millet (*Eleusine coracana* (L.) Gaertn.). Plant Science, 169(4), 657–667. https://doi.org/10.1016/j.plantsci.2005.05.009

Ma, J., Lyu, Y., Liu, X., Jia, X., Cui, F., Wu, X., Deng, S., & Yue, C. (2022). Engineered probiotics. Microbial Cell Factories, 21(1), 72. https://doi.org/10.1186/s12934-022-01799-0

Malini, B., Sunil, C. K., Rawson, A., Vidyalakshmi, R., & Venkatachalapathy, N. (2024). Effect of pineapple core powder on white finger millet vegan probiotic beverage: Nutrition, sensory and storage. Food Chemistry Advances, 4, 100593. https://doi.org/10.1016/j.focha.2023.100593

Markowiak-Kopeć, P., & Śliżewska, K. (2020). The effect of probiotics on the production of short-chain fatty acids by human intestinal microbiome. Nutrients, 12(4), 1107. https://doi.org/10.3390/nu12041107

Marti, A., Cardone, G., & Pagani, M. A. (2020). Sprouted cereal grains and products. In *Innovative processing technologies for healthy grains* (pp. 113–141), John Wiley & Sons, Ltd. https://doi.org/10.1002/9781119470182.ch6

Meena, L., Buvaneswaran, M., Byresh, T. S., Sunil, C. K., Rawson, A., & Venkatachalapathy, N. (2023). Effect of ultrasound treatment on white finger millet-based probiotic beverage. Measurement: Food, 10, 100090. https://doi.org/10.1016/j.meafoo.2023.100090

Meena, L., Malini, B., Byresh, T. S., Sunil, C. K., Rawson, A., & Venkatachalapathy, N. (2024). Ultrasound as a pre-treatment in millet-based probiotic beverage: It's effect on fermentation kinetics and beverage quality. Food Chemistry Advances, 4, 100631. https://doi.org/10.1016/j.focha.2024.100631

Navyashree, N., Buvaneswaran, M., Sunil, C. K., Rawson, A., & Natarajan, V. (2023). Development of white finger millet probiotic beverage using *Lactocaseibacillus rhamnosus* (LGG): Process optimization and study of physicochemical and nutritional properties. Journal of Food Process Engineering, 46(10), e14200. https://doi.org/10.1111/jfpe.14200

Nefale, F. E., & Mashau, M. E. (2018). Effect of germination period on the physicochemical, functional and sensory properties of finger millet flour and porridge. Asian Journal of Applied Sciences, 6(5). https://doi.org/10.24203/AJAS.V6I5.5466

Nie, Y., Luo, F., & Lin, Q. (2018). Dietary nutrition and gut microflora: A promising target for treating diseases. Trends in Food Science & Technology, 75, 72–80. https://doi.org/10.1016/j.tifs.2018.03.002

Nobre, C., Simões, L. S., Gonçalves, D. A., Berni, P., & Teixeira, J. A. (2022). Fructo-oligosaccharides production and the health benefits of prebiotics. In Current developments in biotechnology and bioengineering (pp. 109–138). Elsevier. https://doi.org/10.1016/B978-0-12-823506-5.00002-3

Pais, P., Almeida, V., Yılmaz, M., & Teixeira, M. C. (2020). *Saccharomyces boulardii*: What makes it tick as successful probiotic? Journal of Fungi, 6(2), 78. https://doi.org/10.3390/jof6020078

Palaniappan, A., Balasubramaniam, V. G., & Antony, U. (2017). Prebiotic potential of xylooligosaccharides derived from finger millet seed coat. Food Biotechnology, 31(4), 264–280. https://doi.org/10.1080/08905436.2017.1369433

Precup, G., Pocol, C. B., Teleky, B. E., & Vodnar, D. C. (2022). Awareness, knowledge, and interest about prebiotics—A study among Romanian consumers. International Journal of Environmental Research and Public Health, 19(3), 1208. https://doi.org/10.3390/ijerph19031208

Ramakrishna, C., Singh, S., Raghavendrarao, S., Padaria, J. C., Mohanty, S., Sharma, T. R., & Solanke, A. U. (2018). The membrane tethered transcription factor EcbZIP17 from finger millet promotes plant growth and enhances tolerance to abiotic stresses. Scientific Reports, 8(1), 2148. https://doi.org/10.1038/s41598-018-19766-4

Selladurai, M., Pulivarthi, M. K., Raj, A. S., Iftikhar, M., Prasad, P. V., & Siliveru, K. (2023). Considerations for gluten free foods-pearl and finger millet processing and market demand. Grain & Oil Science and Technology, 6(2), 59–70. https://doi.org/10.1016/j.gaost.2022.11.003

Shivakumar, D., Gunduraj, A., & Yanjarappa Sreenivasa, M. (2023). Isolation, evaluation, and in-vitro characterization of *Lactobacillus* isolated from traditional fermented food finger millet (*Eleusine coracana*) batter for their probiotic attributes. Food Biotechnology, 37(4), 338–363. https://doi.org/10.1080/08905436.2023.2267121

Siddiqui, F. J., Belayneh, G., & Bhutta, Z. A. (2021). Nutrition and diarrheal disease and enteric pathogens. In *Nutrition and infectious diseases: shifting the clinical paradigm* (pp. 219–241). Springer International Publishing. https://doi.org/10.1007/978-3-030-56913-6_8

Singh, R. K., Phanindra, M. L. V., Singh, V. K., Raghavendrarao, S., Solanke, A. U., & Kumar, P. A. (2014). Isolation and characterization of drought responsive EcDehydrin7 gene from finger millet (*Eleusine coracana* (L.) Gaertn.). Indian Journal of Genetics and Plant Breeding, 74(4), 456–462. https://doi.org/10.5958/0975-6906.2014.00870.0

Singh, U. M., Metwal, M., Singh, M., Taj, G., & Kumar, A. (2015). Identification and characterization of calcium transporter gene family in finger millet in relation to grain calcium content. Gene, 566(1), 37–46. https://doi.org/10.1016/j.gene.2015.04.021

Singh, V., Lee, G., Son, H., Amani, S., Baunthiyal, M., & Shin, J. H. (2022). Anti-diabetic prospects of dietary bio-actives of millets and the significance of the gut microbiota: A case of finger millet. Frontiers in Nutrition, 9, 1056445. https://doi.org/10.3389/fnut.2022.1056445

Sneha, K., Kumar, M., Kaushik, D., Kumar, A., & Bansal, V. (2023). To study the germination effect on finger millet, pearl millet and buckwheat: It's impact on phytochemical properties and their prebiotic effect. Journal of Food Chemistry and Nanotechnology, 9(S1), S166–S173. https://doi.org/10.17756/jfcn.2023-s1-022

Sood, S., Joshi, D. C., Chandra, A. K., & Kumar, A. (2019). Phenomics and genomics of finger millet: Current status and future prospects. Planta, 250(3), 731–751. https://doi.org/10.1007/s00425-019-03159-6

Sood, S., Kumar, A., Babu, B. K., Gaur, V. S., Pandey, D., Kant, L., & Pattnayak, A. (2016). Gene discovery and advances in finger millet [*Eleusine coracana* (L.) Gaertn.] genomics—An important nutri-cereal of future. Frontiers in Plant Science, 7, 1634. https://doi.org/10.3389/fpls.2016.01634

Souza, H. F. D., Carosia, M. F., Pinheiro, C., Carvalho, M. V. D., Oliveira, C. A. F. D., & Kamimura, E. S. (2021). On probiotic yeasts in food development: *Saccharomyces boulardii*, a trend. Food Science and Technology, 42, e92321. https://doi.org/10.1590/fst.92321

Tsehaye, Y., Berg, T., Tsegaye, B., & Tanto, T. (2006). Farmers' management of finger millet (Eleusine coracana L.) diversity in Tigray, Ethiopia and implications for on-farm conservation. Biodiversity & Conservation, *15*(13), 4289–4308. https://doi.org/10.1007/s10531-005-3581-3

Vernocchi, P., Del Chierico, F., & Putignani, L. (2020). Gut microbiota metabolism and interaction with food components. International Journal of Molecular Sciences, 21(10), 3688. https://doi.org/10.3390/ijms21103688

Vila-Real, C., Pimenta-Martins, A., Mbugua, S., Hagrétou, S. L., Katina, K., Maina, N. H., Pinto, E., & Gomes, A. M. (2022). Novel synbiotic fermented finger millet-based yoghurt-like beverage: Nutritional, physicochemical, and sensory characterization. Journal of Functional Foods, 99, 105324. https://doi.org/10.1016/j.jff.2022.105324

Wambi, W., Otienno, G., Tumwesigye, W., & Mulumba, J. (2021). Genetic and genomic resources for finger millet improvement: Opportunities for advancing climate-smart agriculture. Journal of Crop Improvement, 35(2), 204–233. https://doi.org/10.1080/15427528.2020.1808133

Yadav, S., Mishra, S., & Pradhan, R. C. (2023). Effect of fermentation with probiotic *Lactiplantibacillus plantarum* (MCC 2974) on quality characteristics and in vitro digestibility of finger millet milk. ACS Food Science & Technology, 4(1), 152–160. https://doi.org/10.1021/acsfoodscitech.3c00441

Yenasew, A., & Urga, K. (2022). Effect of germination period on physiochemical properties of elite finger millet varieties. Cogent Food & Agriculture, 8(1), 2093045. https://doi.org/10.1080/23311932.2022.2093045

Yogeesh, L. N., Naryanareddy, A. B., Nanjareddy, Y. A., & Gowda, M. C. (2016). High temperature tolerant genotypes of finger millet (*Eleusine coracana* L.). Nature Environment and Pollution Technology, 15(4), 1293.

You, S., Ma, Y., Yan, B., Pei, W., Wu, Q., Ding, C., & Huang, C. (2022). The promotion mechanism of prebiotics for probiotics: A review. Frontiers in Nutrition, 9, 1000517. https://doi.org/10.3389/fnut.2022.1000517

10

Finger Millet Food Product Development

Krishna Aayush, Gurvendra Pal Singh, Prafull Chavan, Rahul Agarwal, and Rahul Saxena

10.1 Introduction

Finger millet (*Eleusine coracana*), commonly referred to as ragi, is an ancient grain native to Africa and India that has gained renewed attention in recent years due to its exceptional nutritional profile and versatility in food applications (Saleh et al., 2013). Traditionally consumed in rural communities, it is now recognized globally for its potential as a staple and functional food ingredient. The increased interest in finger millet is driven by several factors, including its health benefits, adaptability to different growing environments, and relevance to contemporary dietary trends (Sathiya & Chithra, 2019). This grain plays an essential role in food security, particularly in arid regions, where it thrives despite harsh climatic conditions and nutrient-deficient soils. The resilience of finger millet, combined with its nutrient-rich composition, makes it an attractive crop in the face of climate change and growing food demands. Its history is deeply intertwined with the customs of these regions, where it has been used in the preparation of staples such as porridges, flatbreads, and fermented beverages. Despite its historical significance, finger millet is now gaining popularity beyond its traditional contexts, largely due to its impressive nutritional composition, which aligns with current consumer demands for healthy and functional foods (Sinha & Sharma, 2022). Additionally, its suitability for gluten-free diets has made it a favorite among those with celiac disease and gluten intolerance, further expanding its appeal (Saleh et al., 2013).

The nutritional profile of finger millet is particularly notable. It is an excellent source of complex carbohydrates, proteins, and dietary fiber, and it boasts a low glycemic index (GI), making it an ideal food for individuals looking to manage blood sugar levels (Jagati & Mahapatra, 2021). Its carbohydrate content, largely in the form of starch, is slowly digested, leading to a gradual release of glucose into the bloodstream, which helps prevent spikes in blood sugar levels. This makes finger millet an appropriate dietary choice for individuals with diabetes or those at risk of developing

 DOI: 10.1201/9781003535119-10

the condition (Jagati & Mahapatra, 2021). Additionally, finger millet is a rich source of essential amino acids such as methionine and lysine, which are often limited in other cereal grains, thus offering a more balanced protein profile (Shobana et al., 2013).

Finger millet is also particularly noted for its high calcium content, making it one of the best non-dairy sources of this crucial mineral. It contains approximately 350 mg of calcium per 100 g, making it an essential food for bone health, particularly in regions where dairy products may not be widely consumed (Antony Ceasar et al., 2018). The grain is also rich in iron, zinc, and magnesium, all of which are essential for maintaining various physiological functions, such as oxygen transport, immune response, and bone development. Moreover, finger millet is an excellent source of antioxidants, including phenolic compounds, which play a role in reducing inflammation and mitigating the risk of chronic diseases such as cardiovascular disease and certain cancers (Antony Ceasar et al., 2018).

When used in product formulation, finger millet can take various forms, including flour and whole grain, each offering unique advantages. As a flour, finger millet provides a fine texture and mild flavor, making it suitable for blending with other flours in the preparation of baked goods, snacks, and traditional dishes. Its gluten-free nature offers a valuable alternative to consumers with gluten intolerance or celiac disease, who often struggle to find nutritious and tasty food options (Jagati & Mahapatra, 2021). In baked goods, finger millet flour can be used to develop cookies, cakes, and muffins that are not only nutritionally dense but also have a distinct earthy flavor that can be enhanced with the addition of spices, nuts, or fruits. For instance, finger millet cookies with cardamom or muffins made with finger millet and dates provide a wholesome snack that is both flavorful and health-promoting.

The incorporation of finger millet into various product types offers numerous possibilities for innovation in the food industry. In snack formulations, finger millet can be used to create nutritious chips, crackers, and puffed snacks that cater to the growing demand for healthier snack options. These products are particularly appealing to consumers looking for convenient yet healthy food choices.

In beverages, finger millet can be utilized to develop both traditional and modern drinks. Traditional Indian beverages such as ragi malt have been consumed for generations, offering a nutritious and refreshing option. Modern adaptations, like finger millet smoothies, are becoming increasingly popular in the health and wellness market. These beverages cater to the demand for plant-based, dairy-free alternatives that are both nutritious and delicious. The inclusion of finger millet in these drinks not only boosts the nutrient content but also adds a distinctive flavor that complements other ingredients like fruits and seeds.

In today's competitive food industry, product development is essential for innovation and staying relevant in the market. As consumer preferences evolve, there is an increasing demand for nutritious, functional, and convenient food products. The incorporation of finger millet into food formulations aligns with several major market trends, including the rising interest in gluten-free products, plant-based diets, and functional foods that offer specific health benefits such as improved digestion, enhanced bone health, and blood sugar management.

Finger millet plays a significant role in addressing nutritional challenges, especially in regions where malnutrition and micronutrient deficiencies are prevalent. Its high content of essential nutrients like calcium, iron, and fiber makes it a valuable ingredient in food products designed to combat these deficiencies.

The importance of this chapter lies in highlighting the potential of finger millet as a versatile and nutritious ingredient for product development. By understanding its nutritional benefits, functional properties, and market appeal, food technologists and product developers can create innovative food products that meet the growing demand for health-oriented foods, ultimately improving consumer well-being and addressing global nutritional challenges.

10.2 Formulation and Ingredient Selection

Formulating food products is a complex process that balances nutritional value, sensory appeal, and functional requirements. At its core, food product formulation aims to create products that are not only nutritious but also appealing in taste, texture, and appearance. Food product formulation is a complex process that involves balancing nutritional needs and sensory appeal to create products that are not only healthy but also enjoyable to consume (Fellows, 2022). The nutritional balance is critical, especially in the context of finger millet (*E. coracana*), which is recognized for its rich nutrient profile, including high levels of dietary fiber, protein, and essential minerals such as calcium and iron. Finger millet's unique composition makes it an excellent source for various food products focusing for nutritional deficiencies, particularly in populations at risk of malnutrition and diet-related chronic diseases. In addition to nutritional balance, sensory appeal plays a significant role in product formulation. This includes taste, texture, aroma, and visual aspects that can influence consumer acceptance (Gaikwad et al., 2024). Finger millet has a distinct nutty flavor and a coarse texture, which can enhance the sensory profile of food products. The use of functional ingredients and additives is also essential in formulation. These may include natural preservatives, flavor enhancers, and texturizers that can improve the product's shelf life and overall quality. For instance, incorporating enzymes can aid in breaking down complex carbohydrates, enhancing digestibility, and improving the nutritional availability of finger millet-based foods (Patil et al., 2023).

Functional ingredients and additives play a critical role in food product formulation. These components are added to food not just for their nutritional benefits but also to improve the texture, stability, shelf-life, and overall quality of the product (Sruthi & Rao, 2021). Functional ingredients include dietary fibers, omega-3 fatty acids, probiotics, and plant-based proteins, which provide health benefits beyond basic nutrition. Additives, such as emulsifiers, stabilizers, preservatives, and colorants, are used to enhance the physical and chemical properties of food, ensuring that it remains fresh, safe, and visually appealing throughout its shelf life (Novais et al., 2022).

10.3 Role of Finger Millets in Product Formulation

10.3.1 Direct Products of Finger Millet

Finger millet is a versatile grain that can be transformed into various products through minimal processing. Different processing techniques to develop various millet based product is shown in Figure 10.1. These direct products

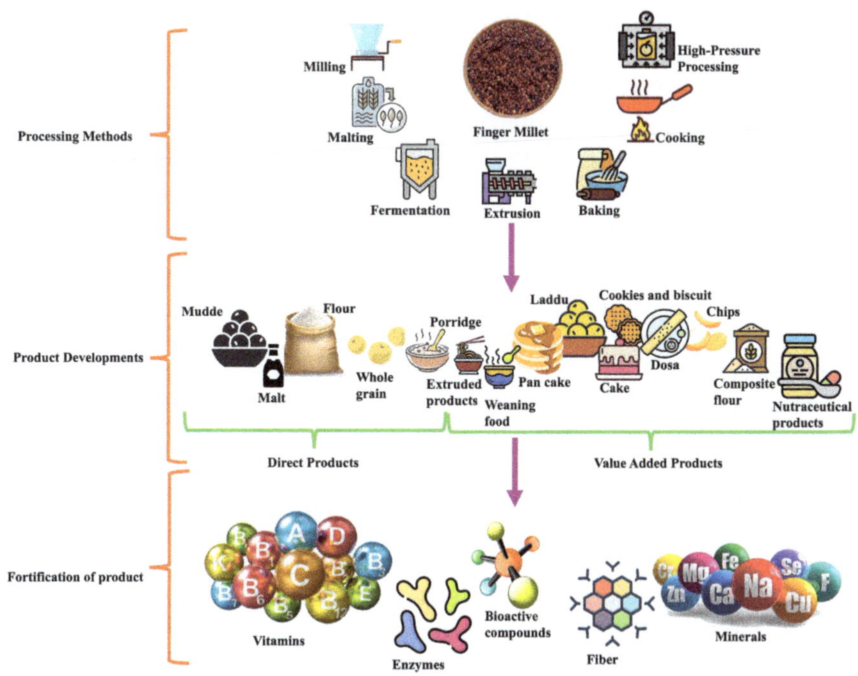

FIGURE 10.1
Different product development and fortification from finger millet by various processing techniques.

retain most of the grain's natural characteristics, ensuring a rich nutritional profile and allowing for diverse culinary applications. Further, explores the formulation, preparation, and nutritional properties of the primary direct products of finger millet.

10.3.1.1 Whole Grains

Whole finger millet grains can be prepared and consumed similarly to other staple grains like rice or quinoa. To prepare, the grains should be thoroughly rinsed to remove any dust or impurities. After rinsing, they can be cooked by boiling in water, with a typical ratio of 1 part finger millet to 2.5 parts water. The grains are simmered over medium heat for about 30–40 minutes or until they become tender. The cooked grains can be served as a base for various dishes, mixed with vegetables, or used in salads (Hall & Rudebjer, 2016). Whole finger millet grains are an excellent source of complex carbohydrates, providing sustained energy release. They are rich in dietary fiber, which aids in digestion and helps regulate blood sugar levels, making them an ideal food for diabetics. The grains are also high in essential minerals such as calcium, magnesium, iron, and potassium, contributing to bone health, muscle function, and overall metabolic processes. Additionally, finger millet contains essential amino acids like methionine, which are often lacking in other cereal grains, making it a valuable protein source in vegetarian and vegan diets (Antony Ceasar et al., 2018). Kharat et al. (2019) did the extrusion processing of finger millet resulted in an expansion ratio of 3.18 and a water absorption index of 3.54 g/g. Finger millet showed lower expansion compared to foxtail and pearl millet but still demonstrated good potential for use in direct expanded extruded foods.

10.3.1.2 Finger Millet Flour

Finger millet flour is made by grinding whole grains into a fine powder. The flour is gluten-free, making it an excellent alternative for individuals with gluten intolerance or celiac disease. It can be used in a variety of culinary applications, including making porridges, pancakes, flatbreads, and as a thickening agent in soups and stews. To make finger millet flour, the grains are cleaned, dried, and ground into a fine powder using a stone mill or electric grinder.

Finger millet flour retains most of the grain's nutritional benefits (Saha et al., 2011). It is high in dietary fiber, which supports digestive health and helps maintain stable blood glucose levels. The flour is rich in calcium, with higher calcium content than most other grains, supporting bone development and preventing osteoporosis. It also contains iron, which is essential to produce hemoglobin and overall oxygen transport in the body. The presence of antioxidants, such as polyphenols and phytates, adds to its health benefits by reducing oxidative stress and inflammation in the body.

10.3.1.3 Porridge

Finger millet porridge is a traditional dish that is simple to prepare and highly nutritious. To make the porridge, finger millet flour is gradually mixed with water or milk to prevent lumps from forming. This mixture is then cooked over medium heat, stirring continuously until it thickens to the desired consistency. The porridge can be flavored with a variety of ingredients, such as fruits, nuts, honey, or jaggery, depending on personal preferences. It is commonly consumed as a breakfast item or as a nutritious snack. Finger millet porridge is rich in carbohydrates, providing a quick source of energy, particularly for children, athletes, and those recovering from illness (Onyango et al., 2022). The use of whole grain flour in the porridge ensures a high fiber content, promoting satiety and aiding in weight management. The addition of milk enhances the protein and calcium content, making the porridge more balanced in terms of macronutrients. When fortified with fruits and nuts, the porridge also provides vitamins, minerals, and healthy fats, further contributing to a balanced diet. In one of the study of Nefale and Mashau (2018) germination of finger millet significantly reduced pH (from 6.24 to 5.72) and increased titratable acidity (from 0.50% to 1.34% lactic acid) after 72 hours. Additionally, the viscosity of porridge decreased drastically from 4777 cP to 420 cP with germination, while oil and water absorption capacities improved significantly.

10.3.1.4 Ragi Balls (Mudde)

Ragi balls, also known as mudde, are a staple food in several rural regions of India, particularly in Karnataka. To prepare ragi balls, finger millet flour is mixed with water in a pot, with continuous stirring to prevent lumps. The mixture is heated until it reaches a dough-like consistency, at which point it is allowed to cool slightly. Once cool enough to handle, small portions of the dough are shaped into balls using wet hands. These balls are typically served with various curries, gravies, or buttermilk (Avadhut et al., 2020). Ragi balls are a nutrient-dense food, providing a substantial amount of energy and nutrients in a small serving. They are high in dietary fiber, which aids digestion and helps control blood sugar levels. The balls are also rich in calcium, which supports bone health, and iron, which is vital for maintaining healthy blood oxygen levels. Due to their high fiber content and low GI, ragi balls are particularly suitable for individuals with diabetes or those looking to manage their weight. The slow digestion of complex carbohydrates in ragi balls provides a steady energy release, keeping individuals full for longer periods. Avadhut et al. (2020) showed in their study that ragi ball was most accepted, showing superior nutritional properties with 10.5% fat, 10.5% protein, 67.45% carbohydrates, and an energy value of 405.9 kcal/100 g.

10.3.1.5 Ragi Malt

Ragi malt is made from sprouted finger millet grains, which are dried and ground into a fine powder. To prepare the malt, the sprouted grains are soaked in water overnight, then drained and allowed to germinate for 2–3 days. Once sprouted, the grains are sun-dried and ground into a powder. This malt powder can be mixed with water, milk, or yogurt to create a smooth beverage. It is often sweetened with honey, jaggery, or sugar and can be flavored with cardamom or other spices. Ragi malt is commonly used as a weaning food for infants due to its high nutritional content and easy digestibility. Sprouting finger millet increases its nutritional value by enhancing the bioavailability of nutrients and reducing anti-nutritional factors such as phytic acid. Ragi malt is particularly high in calcium, making it beneficial for bone health, especially in growing children and the elderly. It is also rich in iron, aiding in the prevention of anemia, and contains essential vitamins like vitamin C (enhanced during sprouting), which helps in iron absorption. The malt is a good source of protein and provides essential amino acids, supporting growth and development in children. Its easy digestibility makes it suitable for people of all ages, including those with compromised digestive systems or recovering from illness (Jagati & Mahapatra, 2021).

10.3.2 Value-Added Products Using Finger Millet

Value addition refers to the process of enhancing the appeal and nutritional value of finger millet through various processing techniques and product development. This can significantly increase its marketability and consumer acceptance. Some notable value-added products include:

10.3.2.1 Ragi Laddoo

Ragi laddoo is a popular Indian sweet made from finger millet flour, jaggery, and ghee, often enriched with nuts and seeds. To prepare ragi laddoo, finger millet flour is first roasted in ghee over a low flame until it turns aromatic and slightly golden. Roasting the flour enhances its flavor and removes any raw taste. Separately, jaggery is melted with a little water to create a syrup, which is then mixed with the roasted flour. Nuts and seeds such as almonds, cashews, and sesame seeds are added for additional crunch and nutrition. The mixture is allowed to cool slightly before being shaped into small, round balls. Ragi laddoos are a nutritious snack or dessert that provides a good source of complex carbohydrates, dietary fiber, and essential minerals such as calcium and iron. The presence of jaggery adds unrefined sugar, which is rich in iron and helps boost hemoglobin levels. Nuts and seeds enhance the laddoos' protein, healthy fats, and micronutrient content, including vitamin E, magnesium, and zinc. The use of ghee not only adds a rich flavor but also provides beneficial fats that aid in the absorption of fat-soluble vitamins. Ragi laddoos are a wholesome treat, especially for children, lactating mothers, and the elderly, due to their nutrient density and energy-boosting properties. Incorporating finger millet into traditional laddoo and dhuska recipes significantly improved their mineral, fiber, calcium, copper, zinc, and β-carotene content. Sensory evaluations showed that the enriched laddoo was "liked very much," while the finger millet-enhanced dhuska was "moderately" liked (Sinha & Sharma, 2022).

10.3.2.2 Ragi Cookies and Biscuits

Ragi cookies and biscuits incorporate finger millet flour into traditional cookie recipes to enhance their nutritional content. To make ragi cookies, finger millet flour is mixed with whole wheat flour or oat flour, butter or oil, and natural sweeteners like honey or jaggery. Baking powder, baking soda, and a pinch of salt are added to the mixture for texture and flavor. The dough is then shaped into cookies and baked at a moderate temperature until golden brown. Variations can include the addition of chocolate chips, nuts, dried fruits, or spices like cinnamon for extra flavor and nutrition. Various studies have demonstrated the potential of ragi and multigrain flour to enhance the nutritional value of bakery products. Jamale et al. (2022) found that cookies made from a blend of refined wheat flour, ragi flour, and whole wheat flour (40:30:30) significantly improved calcium and iron content, with the most acceptable formulation containing 9.46% protein and 150 mg calcium per 100 g. Rana Jawaharlal Nehru Krishi Vishwavidyalaya et al. (2021) highlighted that substituting maida with sprouted ragi millet flour up to 25% increased calcium, phosphorus, and iron content while improving overall acceptability. Suneetha and Rao (2019) reported that cookies enriched with an improved ragi variety (Indravathi) had higher mineral, fiber, calcium, and iron content than locally available ragi cookies. Baghel et al. (2020) emphasized that biscuits made with 28% ragi and 2% Ashwagandha, particularly when ghee was used, scored highest in sensory evaluation and offered better nutritional benefits compared to white butter biscuits. Finally, Shipra et al. (2017) concluded that incorporating multigrain flour (T4 with 40:60 refined flour to multigrain ratio) in sweet biscuits significantly increased protein, calcium, and fiber content, making them nutritionally superior while maintaining reasonable cost-effectiveness. Ragi cookies are marketed as healthy snacks, appealing to health-conscious consumers. They are rich in dietary fiber, which aids in digestion and helps maintain stable blood sugar levels, making them suitable for diabetics and those looking to manage their weight. The cookies also provide a good amount of calcium, essential for bone health, and iron, which supports oxygen transport and prevents anemia. By using natural sweeteners and healthy fats, ragi cookies offer a more nutritious alternative to traditional biscuits, which are often high in refined sugars and trans fats.

10.3.2.3 Ragi Pancakes and Dosas

Finger millet flour can be used to make pancakes or dosas, which are popular breakfast items in many Indian households. For ragi pancakes, the flour is mixed with milk or water, eggs (optional), a pinch of salt, and a leavening agent like baking powder. The batter is poured onto a hot griddle and cooked until bubbles form on the surface, then flipped to cook the other side. Ragi dosas are made with a slightly different batter, combining finger millet flour with fermented rice batter or urad dal batter to create a thin, crispy crepe. The dosas are cooked on a hot tawa (griddle) until golden brown. Ragi pancakes and dosas are nutritious, gluten-free breakfast options that provide complex carbohydrates and dietary fiber, helping to keep you full for longer and maintain energy levels throughout the day. When fortified with vegetables or other grains, they offer a more balanced meal, supplying vitamins, minerals, and antioxidants. Ragi dosas, in particular, benefit from fermentation, which enhances nutrient absorption and improves gut health by introducing beneficial bacteria (Jamale et al., 2022). The high calcium and iron content in these dishes supports bone health and oxygen transport, making them ideal for children, athletes, and anemic individuals. The study by Sathiya and Chithra (2019) found that ragi dosa had the highest calcium content (240 mg per portion) due to the calcium-rich finger millet used. It also recorded a medium GI of 68, making it suitable for health-conscious consumers seeking moderate GI foods.

10.3.2.4 Ragi Cake

Incorporating finger millet flour into cake recipes creates a healthier dessert option. To make a ragi cake, finger millet flour is combined with whole wheat flour, baking powder, baking soda, cocoa powder (for chocolate flavor), and a pinch of salt. Wet ingredients like yogurt, oil, and natural sweeteners such as honey or jaggery are mixed in. The batter is then poured into a greased cake pan and baked in a preheated oven until a toothpick inserted into the center comes out clean. The addition of fruits and nuts can further boost the cake's nutritional value. Desai et al. (2010) has reported the improvement in nutritional quality of cake using malted finger millet with respect to the fiber content and minerals content. Breads are also prepared from millet-based composite flours, finger millet in combination with wheat, proso millet and barnyard millet. Ragi cake provides a good balance of carbohydrates, proteins, and fats, making it a more wholesome dessert compared to traditional cakes that rely on refined flour and sugar. The use of finger millet flour increases the fiber content, aiding digestion and promoting a feeling of fullness. The calcium and iron content in ragi contributes to bone health and prevents anemia, while the use of natural sweeteners and healthy fats reduces the overall glycemic load of the cake. This makes ragi cake a suitable dessert for health-conscious individuals who want to indulge without compromising their dietary goals (Jagati & Mahapatra, 2021).

10.3.2.5 Ragi Chips

Finger millet can be processed into chips, providing a crunchy, healthy snack alternative. To make ragi chips, finger millet flour is mixed with spices, salt, and water to form a dough. The dough is then rolled out thinly and cut into desired shapes, such as triangles or rounds. These pieces are either baked or fried until crisp. Seasonings such as garlic powder, paprika, or herbs can be added to enhance flavor. Ragi chips are a nutritious snack option, offering a higher fiber content compared to conventional potato chips. This fiber helps regulate digestion and supports cardiovascular health by reducing cholesterol levels. The chips are also rich in calcium, which is vital for bone density and strength, and iron, which is necessary for healthy blood and oxygen transport. When baked instead of fried, ragi chips have a lower fat content, making them a healthier alternative for snacking without excessive calories or unhealthy trans fats (Jagati & Mahapatra, 2021).

10.3.2.6 Ragi Malt Beverage

The ragi malt beverage is made from sprouted finger millet and is often consumed as a health drink. To prepare ragi malt, finger millet grains are soaked, sprouted, dried, and ground into a fine powder. The malt powder is mixed with water or milk and boiled while stirring continuously to avoid lumps. Sweeteners such as honey or jaggery and flavorings like cardamom or cocoa can be added. The beverage is typically served warm but can also be chilled. Ragi malt is a nutrient-rich beverage that provides high levels of calcium, iron, and essential amino acids. Sprouting the grains increases the bioavailability of these nutrients and reduces anti-nutritional factors such as phytic acid, which can inhibit mineral absorption. The malt is also rich in fiber, aiding in digestion and providing a sustained energy release, making it an excellent option for breakfast or a midday snack. As a functional food, ragi malt supports bone health, muscle function, and overall metabolic processes, making it suitable for people of all ages (Shobana et al., 2013).

10.3.2.7 Weaning Food

Finger millet malt has long been used as a weaning food for infants, especially in regions where the grain is commonly grown. This nutritious grain is rich in calcium and sulfur-containing amino acids, making it an ideal choice for infant nutrition. To create a weaning food, a composite malt flour made from finger millet, Bengal gram, and green gram can be used. This blend is high in both calcium and protein, providing essential nutrients for growing infants. The preparation of weaning food using finger millet malt involves soaking the composite flour in water at room temperature (about 25°C) for 48 hours, or 38 hours in warmer weather. Soaking initiates germination, which enhances the grains' nutritional value. After soaking, the grains are dried using sun or mechanical drying methods, ensuring temperatures stay below 75°C to maintain milling quality. The dried grains are then roasted at 70–80°C to develop a desirable aroma and flavor. After roasting, the grains are pulverized into a fine, ready-to-eat flour by sieving to remove the husk. To prepare a nutritious beverage called "ragi malt," the malted flour is mixed with skim or whole milk powder, sugar, and flavor. Finger millet malt is particularly beneficial due to its high calcium content, vital for bone development in infants, and its essential amino acids, which make it a good protein source. The inclusion of Bengal gram and green gram further boosts the protein content, making the composite flour an excellent complementary food during weaning. Research has shown that germination improves nutrient availability and reduces anti-nutritional factors, enhancing digestibility and nutritional benefits for infants. The combination of finger millet and legumes provides a balanced mix of carbohydrates, proteins, and essential minerals, supporting infants' nutritional needs during the critical weaning phase (Gaikwad et al., 2024).

10.3.2.8 Composite Flours

Composite flours are made by mixing finger millet flour with other flours, such as wheat, rice, or maize, to create a blend that enhances the nutritional profile of various baked goods. The proportion of finger millet flour in the composite mix can vary depending on the desired texture, flavor, and nutritional outcome. This blend can be used to make bread, pancakes, cakes, and biscuits, offering a more nutritious alternative to products made with refined flours. Composite flours combine the nutritional benefits of finger millet with those of other grains, providing a balanced mix of proteins, carbohydrates, and fats. The inclusion of finger millet increases the calcium and iron content, supports bone health, and prevents anemia. The dietary fiber from finger millet and whole grains aids in digestion and helps regulate blood sugar levels. Composite flours can cater to a wider audience by improving the taste and texture of baked goods while enhancing their overall nutritional value, appealing to both health-conscious consumers and those with specific dietary needs (Mitharwal et al., 2021).

10.3.2.9 Nutritional Supplements

Due to its high calcium and fiber content, finger millet can be processed into nutritional supplements or health powders. These supplements are made by grinding finger millet grains into a fine powder, often combined with other nutrient-dense ingredients like flaxseeds, almonds, or quinoa to boost the nutritional profile. The resulting powder can be packaged as a standalone supplement or blended into smoothies, yogurt, or other foods to increase their nutritional value. Finger millet-based nutritional supplements provide a concentrated source of essential nutrients, including calcium, iron, and magnesium, supporting bone health, blood production, and muscle function. The high fiber content aids digestion and helps maintain healthy blood sugar levels, making these supplements ideal for diabetics and those with digestive concerns. Additionally, the inclusion of bioactive compounds and antioxidants in finger millet contributes to overall health by reducing oxidative stress and supporting immune function. These supplements are versatile and can be easily incorporated into various dietary plans, making them a convenient option for boosting daily nutrient intake (Jagati & Mahapatra, 2021).

10.3.2.10 Extruded Products

Extrusion technology is a widely used method for transforming ingredients into value-added products. It combines shear force with high pressure and temperature to modify the properties of starch, making it ideal for producing a range of food products. Finger millet flour exhibits excellent extrusion characteristics, especially when it has a moisture content of 16–18%. When extruded at barrel temperatures between 100–120°C, finger millet flour demonstrates a good expansion index and results in products with a porous, crunchy, and smooth surface texture. Pasta and noodles, popular convenience foods, can be produced using cold extrusion technology with finger millet. Research has explored various formulations for incorporating finger millet into noodles and pasta. For example, Verma and Patel (2013) reported successful formulations including blends of finger millet, soy, and wheat flour in a 1:4:5 ratio, and another blend using finger millet and wheat flour in a 1:1 ratio. Different combinations can be explored to maintain nutritional value and meet consumer preferences. Additionally, Shukla and Srivastava (2014) found that fortifying noodles with 30–50% finger millet and wheat flour could be beneficial for diabetic patients, as it provides a lower glycemic response compared to conventional noodles.

10.3.3 Enhancing Nutritional and Functional Properties

Finger millet is a highly nutritious cereal grain that has been used as a vehicle for fortification with various micronutrients and bioactive compounds. Its rich nutrient profile, including high levels of dietary fiber, protein, and essential minerals such as calcium and iron, makes it an ideal candidate for fortification to enhance the nutritional value of food products. To further enhance the nutritional and functional properties of food products, fortification with vitamins and minerals is a common strategy. Fortification involves the deliberate addition of essential nutrients that may be lacking in the diet or lost during processing. Several studies have explored the fortification of finger millet-based products with minerals, vitamins, and bioactive compounds to address nutritional deficiencies and promote health benefits. The choice of fortification depends on the target demographic and the intended health benefits. For example, calcium and vitamin D fortification would be ideal for products aimed at children and older adults, while iron and folic acid fortification could target women of reproductive age.

10.3.3.1 Mineral Fortification

Finger millet flour has been effectively used as a vehicle for fortification with minerals such as zinc and iron. The addition of minerals like calcium and iron can significantly boost the nutritional profile of finger millet products, making them particularly beneficial for individuals at risk of osteoporosis or anemia. A study by Tripathi and Platel (2010) found that finger millet flour fortified with zinc oxide or zinc stearate, providing 50 mg of zinc per kg of flour, significantly increased the bioaccessibility of the fortified mineral. The addition of EDTA (Ethylenediaminetetraacetic acid) as a co-fortificant further enhanced the bioaccessibility of zinc by threefold. EDTA is a chelating agent that forms stable complexes with minerals like zinc, preventing their interaction with phytates and thereby improving their absorption in the body. The fortified flours maintained their quality and zinc content during storage for up to 60 days, indicating the stability of the fortification process. In another study, Anitha et al. (2021) reported an improved hemoglobin status in children who were given finger millet-based food fortified with iron. The fortification of finger millet with iron can help address iron deficiency anemia, a common nutritional concern in many parts of the world.

10.3.3.2 Vitamin Fortification

While limited research has been conducted on the direct fortification of finger millet-based products with vitamins, the incorporation of finger millet into various food items can indirectly enhance their vitamin content. Finger millet is a good source of B vitamins, such as thiamine, riboflavin, and niacin, which are essential for energy metabolism and overall health (Jagati & Mahapatra, 2021). In the case of finger millet-based products, fortifying with vitamins such as vitamin D, vitamin B12, and folic acid can help address specific nutritional deficiencies prevalent in certain populations. By incorporating finger millet into products like breads, snacks, and beverages, the vitamin content of these foods can be improved.

10.3.3.3 Bioactive Compound Fortification

Finger millet is rich in bioactive compounds, such as polyphenols and phytochemicals, which have been associated with various health benefits. These compounds possess antioxidant, anti-inflammatory, and anti-diabetic properties (Jagati & Mahapatra, 2021). While direct fortification of finger millet-based products with isolated bioactive compounds has not been extensively studied, the incorporation of finger millet itself can increase the bioactive compound content of the final product. The incorporation of finger millet into snacks and beverages can enhance their bioactive compound profile, potentially providing additional health benefits to consumers.

10.3.3.4 Fiber Fortification

Finger millet is a rich source of dietary fiber, with a fiber content ranging from 15% to 20% (Jagati & Mahapatra, 2021). The high fiber content of finger millet makes it an excellent choice for fortifying food products to increase its fiber content. By incorporating finger millet flour or whole grains into products like bread, cookies, and cereals, the fiber content of these foods can be significantly enhanced, promoting digestive health and providing a feeling of satiety. In conclusion, finger millet-based products offer great potential for fortification with minerals, vitamins, bioactive compounds, and fiber. The fortification of finger millet with zinc and iron has been successfully demonstrated, while the incorporation of finger millet itself can indirectly enhance the nutrient profile of various food products. Further research is needed to explore the direct fortification of finger millet-based products with other micronutrients and bioactive compounds to fully harness its nutritional potential and promote the development of nutrient-dense foods.

10.3.3.5 Enzyme Fortification

The use of enzymes and bioactive compounds is another effective way to enhance the functional properties of food products. Enzymes, such as amylases, proteases, and lipases, are used to modify the texture, flavor, and digestibility of food. In the context of finger millet-based products, enzymes can be employed to break down complex carbohydrates, enhancing the digestibility and reducing the GI of the final product (Afrose Subaitha et al., 2024). This enzymatic modification is particularly useful in developing foods for diabetic patients or those with digestive issues. Bioactive compounds, such as antioxidants, phytochemicals, and probiotics, offer additional health benefits by reducing inflammation, improving gut health, and supporting the immune system. Incorporating these bioactive components into finger millet products can enhance their appeal to health-conscious consumers and provide functional benefits beyond basic nutrition.

10.4 Processing Techniques

10.4.1 Primary Processing of Finger Millets

10.4.1.1 Milling of Finger Millet

Milling is a vital process in transforming finger millet grains into flour, which serves as the foundation for various food products like porridges, flatbreads, and snacks. However, due to the unique characteristics of finger millet such as its small size and hard outer husk milling is more complex than it is for other grains like wheat or rice (Kaimal et al., 2021; Kumar et al., 2018). Before the actual milling process begins, the grains need to be thoroughly cleaned to remove impurities such as dirt, stones, foreign grains and seeds. This cleaning step is essential to ensure product safety, reduce wear on machinery, and produce high-quality flour. After cleaning water is added to the cleaned millet grains to moisten them. This softening process serves two main purposes loosens the bran which means the outer bran of finger millet is quite hard. Moistening the grain helps soften it, making it easier to separate from the inner endosperm during the milling process (Onyango et al., 2022). Prepares for dehusking which means proper conditioning helps in dehusking, making the removal of the outer layer more efficient. Conditioning can vary in time and moisture content depending on the grain type and the end product requirements. For finger millet, it's crucial to control this process because over-moistening can lead to the growth of molds. At the same time, under-conditioning makes dehusking less effective. After conditioning and dehusking, the finger millet grains are ground into flour (Onyango et al., 2022; Verma & Patel, 2013). There are two common types of mills used for grinding. Roller mills use a series of rotating cylinders to break down the grain into flour. Roller milling can produce different fractions of flour, such as coarse or fine, depending on the settings. This method tends to produce flour with a consistent texture and particle size. Hammer mills use high-speed rotating blades to crush the grain. This method is faster and more versatile but can produce flour with varying particle sizes. Grinding is a crucial step as it determines the texture and quality of the final flour. The finer the grind, the smoother the flour, which is ideal for certain products like baked goods. Coarser flour, on the other hand, is preferred for porridges and traditional dishes like *ragi mudde* (Shobana et al., 2013; Tripathi et al., 2022). Milling is not just about reducing the size of the grain but also about optimizing the flour's functional properties. Proper milling techniques help retain essential nutrients like calcium, iron, and dietary fiber, which are abundant in finger millet. Well-milled flour is easier to digest, which is especially important for infant foods and nutritional products. The ability to produce different flour textures through milling allows for the creation of a wide range of food products, from traditional porridges to modern snack food (Anitha et al., 2021; Gaikwad et al., 2024).

10.4.1.2 Malting of Finger Millet

Malting is a biochemical process used to enhance the nutritional profile, flavor, and digestibility of finger millet. It involves controlled germination of the grain, which activates natural enzymes that modify its biochemical composition. This process is particularly beneficial for making finger millet more suitable for infant foods and nutritionally enhanced products due to its improved nutrient availability (Gaikwad et al., 2024; Shobana et al., 2013). The malting process can be divided into three main steps: steeping, germination, and drying/kilning. In the first step in the malting process, where finger millet grains are soaked in water to initiate germination. Water absorption is the grains absorb water, swelling in size and softening their outer husk. Controlled Conditions is the steeping process that usually takes place over 1–2 days. This ensures uniform water uptake and initiates the natural enzymatic processes within the grain. During steeping, the moisture content of the millet increases, which is essential to trigger the biological processes required for germination. Once the grains are sufficiently hydrated, they are spread out in a controlled environment to allow for germination (Anitha et al., 2021; Murungweni et al., 2024). During the germination enzyme activation process, natural enzymes like amylases and proteases are activated. These enzymes play a critical role in breaking down the complex starches present in the millet into simpler sugars, primarily maltose. Nutrient bioavailability in germination also enhances the availability of micronutrients such as calcium, iron, and zinc (Anitha et al., 2021; Bisht et al., 2023). It reduces anti-nutritional factors like phytic acid, which otherwise binds to minerals and reduces their absorption. Increased digestibility in this process the breakdown of starches and proteins makes the grain more digestible, which is particularly beneficial for making weaning foods for infants and for people with dietary sensitivities. The germination process typically lasts for 3–5 days, during which the grains develop small shoots. Monitoring temperature, humidity, and airflow during this stage is essential to avoid mold growth and ensure uniform enzyme activation (Chorfa et al., 2022; Sathiya & Chithra, 2019). Malting significantly improves the nutritional and sensory qualities of finger millet, making it ideal for a range of food products. Malting increases the bioavailability of important nutrients, particularly calcium, iron, and other essential minerals. The reduction of phytic acid and other anti-nutritional factors allows for better absorption of these nutrients. The enzymatic breakdown of starches and proteins during malting makes the millet easier to digest. This is especially important for infants, the elderly, and individuals with compromised digestive systems. Malting adds a pleasant, slightly sweet, and nutty flavor to the grain, making it more appealing in food applications like porridges, baked goods, and beverages (Shobana et al., 2013).

10.4.1.3 Fermentation of Finger Millet

Fermentation is a traditional and widely used primary processing method for finger millet, which improves its nutritional profile, digestibility, flavor, and shelf life. This process relies on the action of beneficial microorganisms, such as lactic acid bacteria (LAB) and yeast, to transform the grain or flour into more palatable and nutrient-rich products. The microbial fermentation of finger millet is generally carried out under anaerobic conditions, meaning without oxygen, which is ideal for the growth of certain microbes (Jan et al., 2022; Shobana et al., 2013). Natural fermentation finger millet grains or flour are mixed with water and allowed to ferment naturally for a period ranging from 12 to 48 hours, depending on the ambient temperature. No external starter cultures are added, allowing the indigenous microbes present in the environment or on the millet itself to initiate fermentation. The key microorganisms involved in this process are LAB like *Lactobacillus* species and various types of yeast. These microbes feed on the natural sugars in the millet and convert them into organic acids, primarily lactic acid and some ethanol (Jan et al., 2022; Nigam & Luke, 2016). pH reduction: As fermentation progresses, the production of lactic acid lowers the pH of the millet mixture, creating an acidic environment. This not only enhances the flavor by imparting a tangy taste but also improves the shelf life by inhibiting the growth of spoilage organisms. Time frame: The duration of fermentation can vary. At warmer temperatures, the process takes 12 to 24 hours, while cooler conditions may extend fermentation to up to 48 hours. Natural fermentation is a widely used method for traditional finger millet foods because it requires minimal inputs and produces flavorful, nutrient-rich products (Kumar et al., 2018).

In controlled fermentation, specific starter cultures are introduced to the millet to ensure consistent results and enhanced control over the fermentation process. Starter cultures are the pre-selected microbial strains chosen for their ability to rapidly ferment millet, producing desirable flavors, textures, and enhanced nutrient profiles. Starter cultures ensure that the fermentation process is reliable, with consistent outcomes in terms of taste, texture, and nutritional benefits. Controlled fermentation minimizes the risk of contamination, reduces fermentation time, and provides more uniform products. It is often used in commercial production to ensure product safety and standardization (Shobana et al., 2013).

The nutritional impact of fermentation of finger millet leads to several nutritional improvements, making it highly beneficial in food product development, particularly for weaning foods, health foods, and functional beverages. Fermentation enhances the bioavailability of essential micronutrients such as iron and zinc. Lactic acid fermentation breaks down phytic acid, an anti-nutrient found in millet, which otherwise binds to minerals and reduces their absorption in the body (Jan et al., 2022; Yousaf et al., 2021). Phytic acid and tannins, common anti-nutritional factors in millet, are significantly reduced during fermentation. This makes the millet more digestible and increases its nutritional value. The microbial activity breaks down complex starches and proteins into simpler forms, improving the digestibility of the grain. Fermented millet products are easier to digest, especially for people with sensitive digestive systems, such as infants or the elderly. Fermentation enhances the sensory qualities of finger millet, giving it a slightly sour taste and a softer, more palatable texture. The production of organic acids and other metabolites contributes to the distinct flavor profile of fermented foods (Mitharwal et al., 2021; Tripathi et al., 2022).

10.4.2 Advanced Processing Technologies

Advanced processing technologies of finger millet involve techniques that go beyond the basic milling, malting, and fermentation to create innovative food products with improved texture, flavor, shelf life, and nutrient retention. These methods make finger millet more versatile and suitable for a wide variety of modern food applications, ranging from snacks to ready-to-eat products. This section covers key technologies like extrusion, baking, cooking, high-pressure processing (HPP), and other novel methods (Afrose Subaitha et al., 2024; Shobana et al., 2013).

i. *Extrusion*: It is a high-temperature, short-time process that transforms finger millet into expanded, puffed, or shaped food products. It involves forcing millet flour or dough through a barrel at high temperature and pressure, followed by sudden expansion when the product exits the extruder. In this process finger millet flour is mixed with water and other ingredients to form a dough, which is then fed into the extruder. Inside the extruder, the dough is subjected to high heat (usually between 100°C to 200°C) and pressure. As the dough is forced through the die (a specific shaped opening), it expands and takes on the shape of the die. Once extruded, the product is dried, cooled, and sometimes flavored (Kharat et al., 2019; Mościcki et al., 2012).

ii. *Baking*: it is another widely used method to process finger millet into various food products, from traditional flatbreads to modern baked goods like cakes, cookies, and bread. In this process, finger millet flour is typically mixed with other ingredients such as water, fat, and leavening agents to form dough or batter. The mixture is then baked at a controlled temperature, allowing for moisture evaporation, dough expansion, and browning reactions (Maillard reaction), which contribute to the flavor and texture of the final product like millet bread, biscuits and cookies, cakes and pastries etc. Millet flour provides an excellent base for creating gluten-free baked goods, catering to consumers with celiac disease or gluten sensitivities. Finger millet is high in calcium, iron, and antioxidants, making it a nutrient-dense option for healthy baked goods. Baking enhances the slightly nutty flavor of finger millet, making it more appealing in savory and sweet product (Baghel et al., 2020; Saha et al., 2011).

iii. *Cooking:* its various forms (boiling, steaming, roasting, etc.), is a fundamental processing method used for finger millet in both traditional and modern applications. Cooking finger millet flour with water or milk results in a thick porridge, which is a staple in many cultures, particularly for infants, elderly, and athletes. Finger millet flour is also used to make traditional flatbreads like *roti*, *bhakri*, and *injera* in different parts of India and Africa. These breads are typically cooked on a flat griddle or baked in an oven. Finger millet can be boiled or steamed whole or in flour form to make simple and nutritious dishes. Cooking is a simple and accessible method for processing finger millet at home or on a larger scale. While some vitamins may be lost during cooking, the overall nutrient profile of finger millet remains high, especially in terms of minerals like calcium and iron (Gaikwad et al., 2024; Mościcki et al., 2012)

iv. *High-Pressure Processing (HPP)*: HPP is a novel, non-thermal technique used to extend the shelf life of finger millet products while preserving their nutritional and sensory qualities. HPP uses high pressure (up to 600 MPa) to inactivate microorganisms and enzymes without the need for high temperatures (Suri et al., 2023). The product is placed in a high-pressure chamber filled with water, and pressure is applied uniformly to the product for a few minutes. This treatment inactivates bacteria, yeast, and mold while retaining the natural flavor, color, and nutrients of the millet product. HPP can be applied to finger millet-based beverages, batters, and porridges to extend their shelf life without altering their flavor or nutrient content. Ready-to-eat meals and snacks made from finger millet can also benefit from HPP, providing a longer shelf life without the use of preservatives. Since HPP is a non-thermal process, heat-sensitive nutrients like vitamins and antioxidants are better preserved. HPP-treated millet products have a significantly extended shelf life while maintaining their freshness and quality. HPP is considered a "clean" processing method, as it uses only water and pressure without the need for additives or chemicals (Bangar et al., 2022; Joshi et al., 2023)

v. *Microwave Processing:* Microwave heating is a rapid, efficient method that can be used to process finger millet for producing ready-to-eat meals, snacks, or even for puffing. Quick and energy-efficient processing that maintains the millet's flavor and nutritional content.

vi. *Infrared heating* method uses radiant energy to heat and cook finger millet products evenly. It can be used in drying or baking applications, improving energy efficiency and reducing processing times (Rao et al., 2021).

vii. *Ultrasound waves* are used to improve the extraction of nutrients, enhance emulsification, or reduce processing time in products like finger millet-based beverages.

In summary, advanced processing technologies like extrusion, baking, cooking, high-pressure processing, and novel methods such as microwave and infrared heating have expanded the versatility of finger millet. These techniques enhance the nutritional value, convenience, and sensory attributes of millet-based foods, making them attractive to modern consumers (Joshi et al., 2023).

10.5 Future Prospects

The future of finger millet-based products holds great possibility for innovation, with opportunities to create health-conscious foods that meet rising consumer demand for gluten-free, nutrient-dense, and sustainable options. Finger millet's rich nutritional profile, packed with calcium, iron, dietary fiber, and antioxidants, makes it an ideal base for snacks, beverages, and fortified foods. Integrating it with superfoods like chia, quinoa, and flaxseeds, or functional ingredients like probiotics and omega-3s, could create multi-functional products targeting gut health, energy, and protein intake. Finger millet's role in sustainable agriculture, being drought-resistant and low-input, positions it as a key crop for climate-resilient farming. Future advancements in eco-friendly processing, sustainable packaging, and waste management, like using millet by-products for animal feed or bioenergy, will further enhance its environmental and commercial viability. Industry-academia partnerships will drive the commercialization of innovative millet products, contributing to a more sustainable global food system.

10.6 Conclusion

Finger millet (*E. coracana*), or ragi, is a highly versatile and nutrient-dense grain that offers significant potential in modern food product formulations. As a gluten-free alternative, finger millet flour caters to consumers with celiac disease and gluten intolerance, while its whole grain form provides a rich source of dietary fiber, essential amino acids, and key micronutrients such as calcium, iron, and zinc. These nutritional benefits make finger millet an attractive ingredient for various food applications, from porridges and breakfast cereals to baked goods and snacks. In addition to its impressive health profile, finger millet's mild flavor and adaptable texture allow it to be incorporated into a wide range of innovative products, including nutritious chips, crackers, and beverages. Its low GI and high fiber content make it particularly appealing to those managing weight or blood sugar levels. As consumer interest in functional and health-oriented foods grows, finger millet's potential in product development continues to expand, positioning it as a valuable ingredient in the food industry's ongoing push for nutritious, gluten-free, and sustainable food solutions.

References

Afrose Subaitha, Z., Priyadarshini, S. R., Yoha, K. S., & Moses, J. A. (2024). Impact of post-harvest processing techniques on the glycemic index of millets. In Food Chemistry Advances (Vol. 4, p. 100636). Elsevier. https://doi.org/10.1016/j.focha.2024.100636

Anitha, S., Kane-Potaka, J., Botha, R., Givens, D. I., Sulaiman, N. L. B., Upadhyay, S., Vetriventhan, M., Tsusaka, T. W., Parasannanavar, D. J., Longvah, T., Rajendran, A., Subramaniam, K., & Bhandari, R. K. (2021). Millets can have a major impact on improving iron status, hemoglobin level, and in reducing iron deficiency anemia–A systematic review and meta-analysis. Frontiers in Nutrition, 8, 725529). https://doi.org/10.3389/fnut.2021.725529

Antony Ceasar, S., Maharajan, T., Ajeesh Krishna, T. P., Ramakrishnan, M., Victor Roch, G., Satish, L., & Ignacimuthu, S. (2018). Finger millet [*Eleusine coracana* (L.) Gaertn.] Improvement: Current status and future interventions of whole genome sequence. Frontiers in Plant Science, 9, 382118. https://doi.org/10.3389/fpls.2018.01054

Avadhut, P., Sachin, M., Paresh, D., & Lal, A. (2020). Development and formulation of ragi balls enriched with peanuts and dates. Journal of Pharmacognosy and Phytochemistry, 9(2), 2411–2415. https://www.phytojournal.com/archives/2020.v9.i2.11238/development-and-formulation-of-ragi-balls-enriched-with-peanuts-and-dates

Baghel, S., Saraugi, S. S., Kumar, A., Soni, D. K., & Rathore, A. (2020). Sensory and physical evaluation of herbal biscuit incorporated with Ashwagandha (*Withania somnifera*) and Ragi (*Eleusine coracana*). International Journal of Chemical Studies, 8(5), 136–142. https://doi.org/10.22271/chemi.2020.v8.i5b.10290

Bangar, S. P., Suri, S., Malakar, S., Sharma, N., & Whiteside, W. S. (2022). Influence of processing techniques on the protein quality of major and minor millet crops: A review. Journal of Food Processing and Preservation, 46(12), e17042.

Bisht, K., Bisht, K., Gudadhe, N. N., Raut, A. A., & Dobhal, N. (2023). Nutritional composition, health benefits, production, processing and marketing of finger millet. Indian Journal of Fertilisers, 19(10), 1036–1046.

Sathiya, V. & Chithra, R. (2019). Nutritive value, glycemic index and glycemic load of selected dosa varieties. International Journal of Health Sciences & Research, 9, 215.

Chorfa, N., Nlandu, H., Belkacemi, K., & Hamoudi, S. (2022). Physical and enzymatic hydrolysis modifications of potato starch granules. Polymers, 14(10), 2027.

Desai, A. D., Kulkarni, S. S., Sahoo, A. K., Ranveer, R. C., & Dandge, P. B. (2010). Effect of supplementation of malted ragi flour on the nutritional and sensorial quality characteristics of cake. Advance Journal of Food Science and Technology, 2(1), 67–71.

Fellows, P. J. (2022). *Food Processing Technology: Principles and Practice*. Woodhead Publishing.

Gaikwad, V., Kaur, J., Rasane, P., Kaur, S., Singh, J., Kumar, A., Kumar, A., Sharma, N., Mehta, C. M., & Patel, A. S. (2024). Nutritional significance of finger millet and its potential for using in functional products. Foods and Raw Materials, 12(1), 110–123. https://doi.org/10.21603/2308-4057-2024-1-593

Hall, R. A., & Rudebjer, P. (2016). Proceedings of the 3rd International Conference on Neglected and Underutilized Species (NUS): For a Food-Secure Africa. Accra, Ghana, 25–27 September 2013.

Jagati, P., & Mahapatra, I. (2021). Finger millet (ragi) as an essential dietary supplement with key health benefits: A review. International Journal of Home Science, 7(2), 94–100. https://doi.org/10.22271/23957476.2021.v7.i2b.1152

Jamale, S., Sahoo, A., Jamdar, J., & Patil, S. (2022). Formulation of ragi and whole wheat flour supplemented nutri cookies. The Pharma Innovation Journal, 101(12), 101–104. www.thepharmajournal.com

Jan, S., Kumar, K., Yadav, A. N., Ahmed, N., Thakur, P., Chauhan, D., & Dhaliwal, H. S. (2022). Effect of diverse fermentation treatments on nutritional composition, bioactive components, and anti-nutritional factors of finger millet (*Eleusine coracana* L.). Journal of Applied Biology and Biotechnology, 10(1), 46–52.

Joshi, T. J., Singh, S. M., & Rao, P. S. (2023). Novel thermal and non-thermal millet processing technologies: Advances and research trends. European Food Research and Technology, 249(5), 1149–1160.

Kaimal, A. M., Mujumdar, A. S., & Thorat, B. N. (2021). Resistant starch from millets: Recent developments and applications in food industries. Trends in Food Science & Technology, 111, 563–580.

Kharat, S., Medina-Meza, I. G., Kowalski, R. J., Hosamani, A., Ramachandra, C. T., Hiregoudar, S., & Ganjyal, G. M. (2019). Extrusion processing characteristics of whole grain flours of select major millets (foxtail, finger, and pearl). Food and Bioproducts Processing, 114, 60–71. https://doi.org/10.1016/J.FBP.2018.07.002

Kumar, A., Tomer, V., Kaur, A., Kumar, V., & Gupta, K. (2018). Millets: A solution to agrarian and nutritional challenges. Agriculture & Food Security, 7(1), 1–15.

Mitharwal, S., Kumar, S., & Chauhan, K. (2021). Nutritional, polyphenolic composition and in vitro digestibility of finger millet (*Eleusine coracana* L.) with its potential food applications: A review. In Food Bioscience (Vol. 44, p. 101382). Elsevier. https://doi.org/10.1016/j.fbio.2021.101382

Mościcki, L., Mitrus, M., Wójtowicz, A., Oniszczuk, T., Rejak, A., & Janssen, L. (2012). Application of extrusion-cooking for processing of thermoplastic starch (TPS). Food Research International, 47(2), 291–299.

Murungweni, K. T., Ramashia, S. E., & Mashau, M. E. (2024). Effect of malting on physicochemical, antioxidant, and microstructural properties of finger millet (*Eleusine coracana*) flours. Food Science & Nutrition, 12(1), 547–563.

Nefale, F. E., & Mashau, M. E. (2018). Effect of germination period on the physicochemical, functional and sensory properties of finger millet flour and porridge. Asian Journal of Applied Sciences, 6(5). https://doi.org/10.24203/AJAS.V6I5.5466

Nigam, P. S., & Luke, J. S. (2016). Food additives: Production of microbial pigments and their antioxidant properties. Current Opinion in Food Science, 7, 93–100. https://doi.org/10.1016/j.cofs.2016.02.004

Novais, C., Molina, A. K., Abreu, R. M., Santo-Buelga, C., Ferreira, I. C., Pereira, C., & Barros, L. (2022). Natural food colorants and preservatives: A review, a demand, and a challenge. Journal of Agricultural and Food Chemistry, 70(9), 2789–2805.

Onyango, C., Luvitaa, S. K., Lagat, K., Hüsken, A., Smit, I., & Schmidt, M. (2022). Utilisation of amaranth and finger millet as ingredients in wheat dough and bread for increased agro-food biodiversity. Foods, 11(7), 911.

Patil, P., Singh, S. P., & Patel, P. (2023). Functional properties and health benefits of finger millet (*Eleusine coracana* L.): A review. The Journal of Phytopharmacology, 12(3), 196–202.

Rana Jawaharlal Nehru Krishi Vishwavidyalaya, G. K., Prasad Mishra Mahatma Gandhi Chitrakoot Gramodaya Vishwavidyalaya, S., Rana, G., Mishra, S., Duggal, A., Shukla, S., Singh, N., & Rahangdale, H. (2021). Proximates and sensoric attributes of sprouted ragi flour (SRF) supplemented cookies. The Pharma Innovation Journal, 10(10). https://www.thepharmajournal.com

Rao, M. V., Akhil, K. G., Sunil, C. K., Venkatachalapathy, N., & Jaganmohan, R. (2021). Effect of microwave treatment on physical and functional properties of foxtail millet flour. International Journal of Chemical Studies, 9(1), 2762–2767.

Saha, S., Gupta, A., Singh, S. R. K., Bharti, N., Singh, K. P., Mahajan, V., & Gupta, H. S. (2011). Compositional and varietal influence of finger millet flour on rheological properties of dough and quality of biscuit. LWT, 44(3), 616–621. https://doi.org/10.1016/j.lwt.2010.08.009

Saleh, A. S. M., Zhang, Q., Chen, J., & Shen, Q. (2013). Millet grains: Nutritional quality, processing, and potential health benefits. Comprehensive Reviews in Food Science and Food Safety, 12(3), 281–295. https://doi.org/10.1111/1541-4337.12012

Shipra, S., Virginia, P., & Richa, S. (2017). Chemical science review and letters sensory acceptability, nutritional composition and cost of multigrain sweet biscuits. Chemical Science Review and Letters, 6(24), 2320–2323.

Shobana, S., Krishnaswamy, K., Sudha, V., Malleshi, N. G., Anjana, R. M., Palaniappan, L., & Mohan, V. (2013). Finger millet (ragi, *Eleusine coracana* L.). A review of its nutritional properties, processing, and plausible health benefits. In Advances in Food and Nutrition Research (Vol. 69, pp. 1–39). Academic Press. https://doi.org/10.1016/B978-0-12-410540-9.00001-6

Shukla, K., & Srivastava, S. (2014). Evaluation of finger millet incorporated noodles for nutritive value and glycemic index. Journal of Food Science and Technology, 51(3), 527–534. https://doi.org/10.1007/s13197-011-0530-x

Sinha, R., & Sharma, B. (2022). Utilization of finger millet in traditional recipes of tribal for household nutritional security in Jharkhand. Bhartiya Krishi Anusandhan Patrika. https://doi.org/10.18805/bkap513

Sruthi, N. U., & Rao, P. S. (2021). Effect of processing on storage stability of millet flour: A review. Trends in Food Science & Technology, 112, 58–74.

Suneetha, D. B., & Rao, D. G. P. (2019). Evaluation of finger millet cookies for nutritional and sensory characteristics. Journal of Pharmacognosy and Phytochemistry, 8(3), 2017–2019.

Suri, S., Balasaheb, K. S., Yadav, D. K., Malakar, S., Choudhary, P., Mohapatra, A., & Dhurve, P. (2023). Overview of Millet proteins: Quality characteristics, effect of thermal/non-thermal processing and applications. Food Bioscience, 103434.

Tripathi, B., & Platel, K. (2010). Finger millet (*Eleucine coracana*) flour as a vehicle for fortification with zinc. Journal of Trace Elements in Medicine and Biology: Organ of the Society for Minerals and Trace Elements (GMS), 24(1), 46–51.

Tripathi, M. K., Kumar, A., Mohapatra, D., Jadam, R. S., Selvan, S. S., & Nickhil, C. (2022). Economic, nutritional, and health importance of finger millet. In The Finger Millet Genome (pp. 13–33). Springer.

Verma, V., & Patel, S. (2013). Value added products from nutri-cereals: Finger millet (*Eleusine coracana*). Emirates Journal of Food and Agriculture, 25(3), 169–176. DOI: 10.9755/ejfa.v25i3.10764

Yousaf, L., Hou, D., Liaqat, H., & Shen, Q. (2021). Millet: A review of its nutritional and functional changes during processing. Food Research International, 142, 110197.

11

Advancements in Millet Processing and Value Addition

Amit Verma, Prince, Rinku, and Indu

11.1 Introduction

Agriculture crop production is severely impacted by climate change and global warming. With the increasing population, the demands for food grains and food products are also increasing. But with changing weather crop yield is decreasing. Mostly wheat, barley, rice, and maize are grown which require a lot of water to grow. Considering the decreasing level of water, attention should be paid to underutilized crop such as millet. Millet crop have a lot of varieties such as finger millet, little millet, kudo millet and foxtail millet. In India, ragi and mandua are commonly used names of the finger millet. The origin of finger millet traces back to Ethiopia but it is also grown in India as well as Africa, and is staple food of the locals (Chandra *et al.*, 2016). India produces about 173 lakh tons of millet, which was about 80% of Asia's and 20% of world production. Average global yield was 1239 Kg/ha and the average yield in India was 1239 Kg/ha. In India, in the year 2020–2021 the total production of finger millet was about 19.6 lakh tons (PIB). Finger millet contains protein, fat, dietary fibre, and water-soluble vitamins given in Tables 11.1 and 11.2 respectively. A large number of health advantages are offered by finger millet and the polyphenol content of finger millet is responsible for that. Finger millet helps in losing weight, controlling blood sugar, lowering blood cholesterol, and helps in condition of anemia (Rao *et al.*, 2017).

Food scientists, legislators and nutritionists are now paying more attention to finger millet because of its socioeconomic benefits. The various processing methods and storage environments have great impact on the quality of the finger millet as well as various products made from it. Finger millets contain various sulfur containing essential amino acids like lysine and methionine, which helps in preventing the skin aging. Magnesium is also found in finger millet in good amount which helps in controlling the flow of blood and preventing loss of hairs (Verma and Patel, 2013). Finger

DOI: 10.1201/9781003535119-11

TABLE 11.1

Proximate Composition (per 100 g) (Rao *et al.*, 2017)

Component	Amount (g/100 g)
Moisture	10.89 ± 0.61
Protein	7.16 ± 0.63
Ash	2.04 ± 0.34
Total Fat	1.92 ± 0.14
Dietary Fiber	11.18 ± 1.14
Carbohydrate	66.82 ± 0.73
Energy	1342 ± 10 KJ

Source: Rao *et al.* (2017).

millet also contains high calcium content which contributes to the growth of bones and prevents osteoporosis with and additional advantage of helping in preventing kidney issues (Cormick and Belizán, 2019). Millet and its products have a very limited shelf life which is restricted by high enzymatic activity and their high lipid content. The four main enzyme behind the rancidity are polyphenol oxidase, lipase, peroxidase, and lipoxygenase (Rani *et al.*, 2018). Lipase hydrolyzes the lipid to produce glycerol and fatty acids. Peroxidase and lipoxygenase are the enzymes which further oxidizes the free fatty acids which results in the production of various carbonyl compounds and hydroperoxy fatty acids, which gives unpleasant smell (Taylor and Kruger, 2016). A large number of food preservation techniques are explored to increase the shelf life of millet products along with retaining the nutritional value. Various non-thermal processing techniques like ultra-sonication, irradiation, pulse light treatment, and thermal techniques like ohmic heating, microwave, radiofrequency, and infrared, (given in Figure 11.1) help in preserving millets and their products.

TABLE 11.2

Water Soluble Vitamins Profile of Finger Millet

Component	Value
Thiamine-B_1 (mg)	00.37 ± 0.041
Riboflavin-B_2 (mg)	00.17 ± 0.008
Niacin-B_3 (mg)	01.34 ± 0.020
Pantothenic acid-B_5 (mg)	00.29 ± 0.190
Total-B_6 (mg)	00.05 ± 0.007
Biotin-B_7 (μg)	00.88 ± 0.050
Total folates-B_9 (μg)	34.66 ± 4.970

Source: Indian food composition tables, NIN 2017.

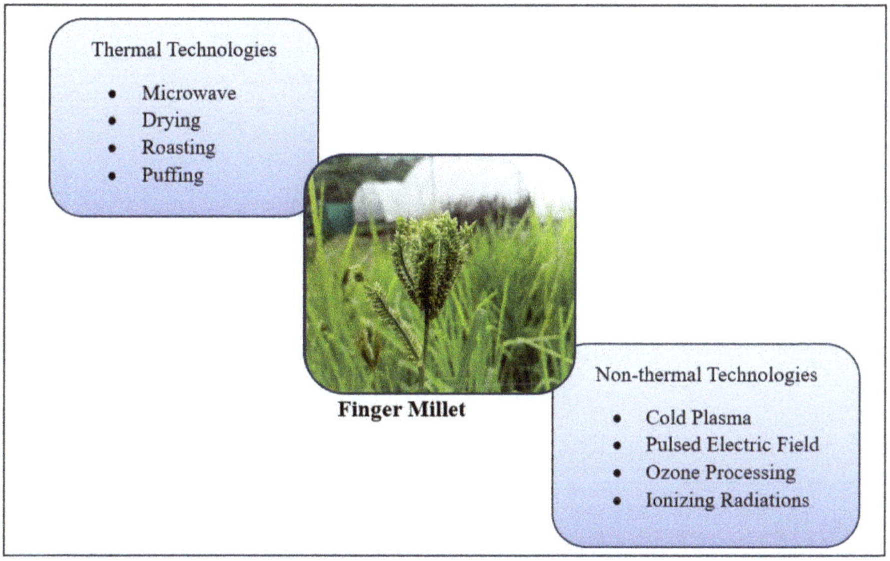

FIGURE 11.1
Advanced technologies used for processing of finger millet.

11.2 Non-Thermal Technologies

Finger millet is considered a nutritious and beneficial food due to its high nutritional value. It contains proteins, amino acids, fibers, carbohydrates, minerals, and various essential nutrients. It can give several effects beneficial to health like antioxidant capacity, antimicrobial and enzyme inhibitory properties due to its high dietary nutrition content. Despite of several advantages of consuming the finger millet, it is still largely consumed in rural areas only (Somarajan and Morya, 2022). In urban areas the consumption of finger millet is limited because of its dark brown in color which is not so aesthetic so the concept of processing of finger millet came to increase the consumption in the urban area as well.

Crop processing is defined as the combination of some physical and mechanical processes by which the raw agricultural products can be converted into final food products or industrial raw materials, which are ready to cook or eat. These processing techniques are utilized to transform raw finger millet grains into a more palatable and higher-quality form. Through primary processing methods, small finger millet grains can be transformed into a variety of products like sprouts, salted snacks, flakes and fermented products. In Indian household the processing techniques which are commonly used for finger millet processing are cooking, soaking, fermenting and sprouting. Soaking and/or cooking are one of the various processing

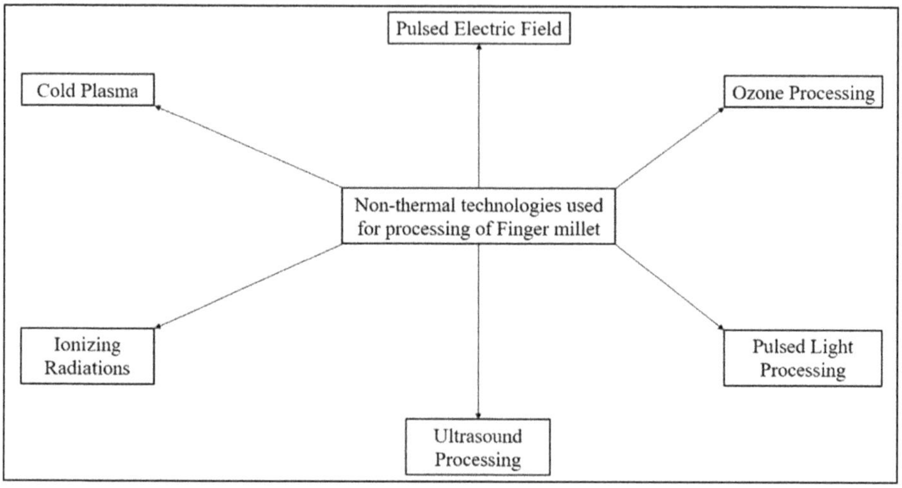

FIGURE 11.2
Non-thermal technologies used for finger millet processing.

methods which are commonly employed to eliminate the effects of anti-nutritional factors present in foods. This chapter outlines different processing techniques used for finger millets. Traditionally, thermal methods are the predominant approach for processing finger millet. Thermal treatments not only inhibit enzymes like lipases and lipoxygenase, thereby prolonging the shelf life of grains, but also impact protein denaturation, aroma, flavor modification, and starch granular structure (Gomez and Martinez, 2016). Structural and compositional changes occur in finger millet when subjected to heat treatment during thermal processing (Joshi *et al.*, 2023), highlighting the importance of utilizing various technologies which do not lead to increase in temperature, that is, non-thermal technologies like pulsed light treatment, irradiation and ozonization. In this chapter, various thermal and non-thermal methods (given in Figure 11.2) used for processing of finger millet are discussed. Various non-thermal technologies, their advantages and disadvantages are tabulated in Table 11.3.

11.2.1 Cold Plasma (CP)

The application of electric energy to gas leads to formation of Cold plasma, which leads to ionization of the gas and the production of highly reactive molecules or particles. The collision of electrons generates the active plasma species (Lokeswari *et al.*, 2021) and disinfects the product, that is, finger millets in this case. It is among the most used non-thermal technology for processing of finger millet. It is named so since during the processing/treatment the temperature of the product is not increased. It is usually performed at ambient or room temperature. Cold plasma can be generated at both atmospheric

TABLE 11.3

Comparison of Various Non-Thermal Technologies Used for Processing of Finger Millet

Non-Thermal Technology	Advantages	Disadvantages
Cold Plasma	Any type of chemical residues are not left behind after processing which makes it non-toxic and environment friendly.	This treatment can be used for the decontamination of the surface only.
High Pressure Processing	Vitamins, flavor compounds and pigments are not affected significantly so the quality of food after processing is comparable to that of fresh food products.	Initial cost of equipment is very high, maintenance cost is also very high.
Ionizing Radiations	These can be utilized for surface irradiation such as electron beams, as well as for processing bulk food such as gamma rays.	Free radicals have indirect adverse impact on the quality of the food product to be processed.
Ozone Processing	Removal of harmful chemical residues without impacting flavor and color of the products.	High cost of equipment due to which not suitable for small processers and different microorganisms have different resistivity for ozone.
Pulsed Electric Field	Deactivates enzymes and pathogens, leading to a longer shelf life while preserving the nutritional and sensory properties.	It results in formation of free radicals in the product processed, which further can take part in various reactions like lipid oxidation etc.
Pulsed Light Processing	Decontamination of surface, microbial inactivation and inactivation of various enzymes like lipases can be effectively achieved.	Efficiency is determined by the composition of food product, microbial activity, and level of contamination in the food material.
Ultrasound Processing	This technique offers minimal processing of the food product with maximum quality and safety.	Heat is produced at higher ultrasound amplitudes, which raises the temperature of the product.

and low pressures by utilizing radio frequency, microwaves, and pulsed electric currents as energy sources (Sirohi *et al.*, 2021). Any type of chemical residues are not left behind in the product after treatment with cold plasma, which makes it non-toxic and environment friendly. Cold plasma treatment is commonly employed for surface decontamination and has the added benefit of enhancing functional properties, improving enzymatic activity, and boosting antioxidant properties (Sarangapani *et al.*, 2016). Studies have demonstrated that cold plasma treatment effectively reduces microbial load on finger millet grains, thereby enhancing safety and extending shelf-life.

Moreover, this method preserves the grain's nutritional quality, including its protein and antioxidant content (Ramakrishnan *et al.*, 2023). In a study conducted by Arsha *et al.* (2025), under the optimized conditions (20 kV, 20 min, 9 cm), CP treatment effectively decreased the total tannin and phytic acid content (anti-nutritional factors). Specifically, the total tannin content was reduced by 27% while phytic acid content decreased by 43%. Additionally, functional properties improved significantly, including increase in water absorption capacity by 12.61%, rise in oil absorption capacity from 0.277 to 1.004 g/g, change in water solubility index by 12.05%. Emulsifying capacity increases by 39% and foaming capacity from 9.31% to 11.62%, coupled with modifications in pasting properties. Overall, CP treatment effectively reduced anti-nutrients and enhanced the functional, morphological, and pasting properties of finger millet flour, suggesting its potential for product development.

11.2.2 High-Pressure Processing

High pressure processing (HPP) is a technique in which the food product to be processed is compressed under very high pressure. In this technique the destruction or inactivation of micro-organisms takes place due to application of high pressure rather than heat, making it a non-thermal technique. It involves applying intense pressure (100–1000 MPa) at ambient temperature (below 40°C) cold or moderate temperatures (<45°C), which enables preservation of flavor and color of the food product after the processing. High pressure is very effective in killing micro-organisms, on the other hand, disruption of covalent bonds of various constituents of food does not takes place due to which the nutrition of the food is maintained. It is a method which helps in maintaining the food quality without the need for excessive heat or chemical preservatives. It can also modify the sensory as well as functional properties of foods, particularly proteins, in a beneficial way. High pressure can denature proteins without using elevated temperature, resulting in increased viscosity and opacity while preserving the fresh flavor. Unlike thermal processes, the high pressure has a very little effect on vitamins, flavors, and pigments. Due to which, the foods processed with high pressure maintain a quality similar to fresh products. HPP treatment can enhance the nutritional profile of finger millet by increasing the bioavailability of essential nutrients and reducing anti nutritional factors. It also improves shelf-life and functional properties of finger millet-based products (Ramakrishnan *et al.*, 2023). In a study conducted by Angioloni and Collar (2012) finger millet flour was subjected to HPP under a pressure of 350 MPa for 10 min at a temperature of 20°C. Their findings included that there is no significant difference in physicochemical characteristics and improved nutritional/sensory profiles of HPP-treated breads so they concluded that the improved digestibility of proteins is attributed to its conformational changes through HPP treatment and inactivation of protease inhibitors.

11.2.3 Ionizing Radiations

Food irradiation is a technique used to preserve food without the use of heat. The food is treated with radiation during the procedure of irradiation, which can incorporate either non-ionizing or ionizing radiations. X-rays, gamma rays and accelerated electron beam are some examples of ionizing radiations. Ionizing radiation has short wavelengths that are intense enough to break atoms (Premi and Khan, 2018). Lipids and carbohydrates are negatively affected by ionizing radiation. Contrarily, an indirect impact occurs when the food itself or the cell components encounter reactive species and free radicals coming from water radiolysis (Bisht *et al.*, 2021). Gamma rays have high penetration rates, so they are often utilized in treating multiplicated foods, while electron beams are primarily used for surface irradiation. Gamma radiation typically kills microbes by disrupting covalent bonds present in their DNA. The radiation is produced by two radioactive Cs-137 and Co-60 isotopes. Irradiation methods have great potential for increasing the shelf life and sterilizing various food products like finger millet grains. Gamma radiation can reduce fungal infections and free fatty acids in finger millets, leading to a decrease in lipase levels. This decline in free fatty acids helps stabilize the flour and extend its shelf life (Mahmoud *et al.*, 2015). The moisture content of finger millet flour can be reduced by treating the flour with gamma radiation. Irradiation effectively reduces microbial contamination in finger millet without adversely affecting its nutritional components. However, it's crucial to optimize dosage to prevent potential alterations in sensory attributes (Ramakrishnan *et al.*, 2023). In a study conducted by Reddy and Viswanath (2019), finger millet grain were subjected to gamma irradiation at doses 2, 5, 10, and 15 kGy. Their findings included that the after processing there is increased protein content (6.85 g/100 g and 8.51 g/100 g in control and 15 kGy irradiated sample) and decreased moisture (12.96 g/100 g and 8.92 g/100 g in control and 15 kGy irradiated sample). Increased hydration capability of the millet protein fractions exposed to gamma ray increased its moisture content. Dissociation of complex proteins by gamma irradiation increased its protein content.

11.2.4 Ozone Processing

Ozone is widely used in food industry for preservation of various food products. Ozone treatment is a non-thermal technology that is environmentally friendly, leaving no residues in the final product. When oxygen molecule reacts with a free radical of oxygen ozone is formed. Ozone treatment can also be used for sanitation purpose, microbes' inhibition, depletion of biomass, grain storage and in beverage processing. When ozone comes in contact with the cell membrane of the micro-organisms. Ozone molecule oxidizes the glycolipids and glycoproteins of the cell membrane. Ozone oxidizes the lipids bound in the membrane and enzymes. Due to which the cell membrane disintegrates, which leads in leaking of the constituents of the cell and cell lysis occurs (Xue *et al.*, 2023). Ozone treatment can reduce microbial load and degrade certain anti-nutritional factors in finger millet. However, excessive exposure may lead to oxidation of lipids and other sensitive compounds, potentially affecting flavor and nutritional quality (Ramakrishnan *et al.*, 2023). In a study conducted by Hayes *et al.* (2020), it was found that finger millet was not sensitive to increasing ozone concentrations and there was some evidence of an increase in weight of individual seed heads with increasing ozone for finger millet.

11.2.5 Pulsed Electric Field (PEF)

This technology involves applying bursts of electric field pulses of high-intensity ranging from 10 to 80 kV/cm between two electrodes to the food placed in the treatment chamber for fractions of second. In other words, there are two electrodes, that is, cathode and anode, the food to be processed is kept in between them and then a potential difference is created between the electrodes due to which electric field pulses are created through the food material to be processed. The peak time in the treatment is increased by employing large number of electric field pulses in series. This technology deactivates enzymes and pathogens, leading to an increased shelf life of the product as well as preserving its nutritional value and taste (Vanga *et al.*, 2021). PEF has been reported to enhance the extraction of bioactive compounds in cereals, potentially improving the nutritional and functional qualities of finger millet-based products (Mollakhalili-Meybodi *et al.*, 2022). In a study conducted by Zhang *et al.* (2023), it was found that the PEF treatment changed the spatial conformation of finger millet, increased its hydrophobicity and solubility in alcohol solution, but did not change the subunit composition distribution. It also promoted the increase in soluble aggregates, resulting in an increase in protein particle size. The α-helix content of the protein decreased, the β-sheet content increased, the number of disulfide bonds increased, and the protein structure became more orderly after treatment. The results showed that the structure and physicochemical properties of finger millet significantly changed after the moderate-intensity PEF treatment.

11.2.6 Pulsed Light (PL) Processing

This technology involves the use of light pulses with wide range of wavelengths for a short duration (Barba *et al.*, 2017). In this treatment method a spectrum of high power electric pulses similar to that of sun (100–200 nm) is made by utilizing inert gas lamps, encompassing infrared, ultraviolet, and visible light (Kramer and Muranyi, 2013). Pulsed light interacts with various constituents of food which leads to increased shelf life and quality of the food product. These changes are brought about by photo-thermal, photo-chemical, and photo-physical processes. The fluence of the pulsed light treatment is a key factor in determining its efficacy (Mahendran *et al.*, 2019). Fluence is the key factor in determining the effectiveness of pulsed light therapy. The fluence value of pulsed light treatment is described in J/cm^2. Pulsed light treatment is majorly used for decontamination of the surface, inactivation of micro-organisms and enzymatic inactivation (Mahendran *et al.*, 2019). Inactivation of lipase in food items can be achieved by using pulsed light treatment which is attributed to the photo-thermal and photo-chemical mechanisms (Jeon *et al.*, 2019). Pulsed light can effectively reduce microbial contamination on finger millet grains, enhancing safety. The rapid treatment minimizes thermal effects, preserving the grain's nutritional and sensory qualities (Ramakrishnan *et al.*, 2023). Pulsed light processing is used for microbial decontamination for effective reduction of surface pathogens and spoilage microorganisms, leading to improved safety and extended shelf life. Pulsed light processing is also helpful for nutritional and sensory preservation as a non-thermal method, PL processing minimizes heat-induced degradation, helping to maintain the nutritional value and sensory attributes of finger millet. A study evaluated the bactericidal effects of two types of pilot-scale intense pulsed light devices on millet. The results demonstrated that PL treatment at a total fluence of 54.43 J/cm^2 achieved a microbial reduction of up to 0.66 log CFU/g. This indicates that PL processing can effectively reduce surface microbial load, potentially enhancing the safety and shelf life of millet grains (Hwang *et al.*, 2018).

11.2.7 Ultrasound Processing

Ultrasound waves are the waves which have a frequency range of 20–100 kHz (Awad *et al.*, 2012). Ultrasound processing is one of the recent technologies that offers the advantage of minimal processing of the food product with maximum safety and quality. Ultrasound waves, are the high-energy waves, which creates alternating compression and rarefaction cycles when passed through food material. This disrupts liquid molecules and modify the food matrix. During cavitation, microbial cell membranes are disrupted and enzymes are inactivated (Cui and Zhu, 2020).

Ultrasound treatment of millet protein concentrates (MPCs) result in higher solubility compared to untreated samples (Nazari *et al.*, 2018). The ultrasound treatment exposes hydrophilic side chains of amino acids of proteins to the water, thereby increasing solubility of the MPC. Additionally, ultrasound treatment improves the emulsification capacity of various food proteins (Sullivan *et al.*, 2016). Ultrasound treatment of finger millet leads to hydration which reduced the anti-nutrients of the finger millets (Yadav *et al.*, 2021). The moisture content of finger millets increased with higher amplitude and longer duration of ultrasound treatment. In contrast, the total phenolic content of finger millet decreased with increased ultrasound amplitude and time (Yadav *et al.*, 2021). Increased time and amplitude lead to higher cavitation in the cellular structure, resulting in greater penetration of the solvent and a decrease in phenolic components (Zhu and Li, 2019). Additionally, ultrasound treatment decreases the phytate content in finger millet (Yadav *et al.*, 2021). Higher ultrasound amplitudes generate heat, increasing the temperature of the finger millet which leads to reduction of phytate (Diouf *et al.*, 2019). Application of ultra-sonication to finger millet has been found to improve its protein digestibility and modify its starch properties, leading to enhanced textural attributes in processed products. This technique also aids in reducing anti nutritional factors, thereby improving the overall nutritional profile (Ramakrishnan *et al.*, 2023). In a study conducted by Nazari *et al.* (2018) finger MPC dispersion (10% w/w) was sonicated for 5, 12.5, and 20 min with an amplitude of 20%, 60%, and 100% with constant pulse durations. They observed that the solubility of MPC has not been significantly influenced by increasing the US time from 5 to 20 min at 18.4 W/cm^2 intensity. At 73.95 W/cm^2 intensity, the solubility has been decreased by increasing the US time from 12.5 to 20 min. A decrease in foaming capacity and no improvement in foaming stability at low intensities and times of US treatment have been observed. They concluded that the conformational changes through ultrasonic treatment affect its surface hydrophobicity.

11.3 Thermal Processing Technologies in Millet Processing

Thermal processing technologies are crucial in millet processing, offering methods to extend shelf life, enhance flavor, and produce new value-added products. The main thermal processing techniques used in millet processing include drying, roasting, puffing, and extrusion. Various thermal technologies, their advantages and disadvantages are tabulated in Table 11.4. Each of these techniques is explored in detail below.

11.3.1 Drying

Drying is a crucial step in millet processing as it helps to reduce the percentage of moisture in grains. This step is crucial to prevent survival of pathogens and enzymes that can cause the millets to spoil. While traditional sun drying has been widely used, it has limitations, including the risk of contamination, uneven drying, and dependence on weather conditions. Modern drying methods offer better control and efficiency.

11.3.1.1 Hot Air Drying

This technique entails circulating hot air around the millet grains, causing moisture to evaporate. Hot air drying is beneficial because it is simple and cost-effective method to maintain the nutrition of millets. Various studies shows that hot air drying can retain various important nutrients like fiber and proteins, while effectively reducing moisture content (Kabeer *et al.*, 2023).

TABLE 11.4

Thermal Technologies and Their Process Parameters for Finger Millet

Thermal Technology	Advantages	Disadvantages	Processing Conditions	References
Roasting	Enhances flavor, reduces anti-nutritional factors, improves digestibility	Risk of over-roasting, loss of some heat-sensitive vitamins	At 112°C–120°C for 15–20 min	Oluwole *et al.* (2019)
Drying	Preserves grain for long-term storage, prevents microbial growth	Time-consuming, may cause slight nutrient loss if temperature is too high	At 50°C–53°C and air velocity 1.5 m/s	Kabeer *et al.* (2023)
Steam Blanching	Quick, effective for enzyme inactivation, preserves color	Partial loss of water-soluble vitamins	At 100°C for 20–30 min	Okolo and Adejumo (2018)
Extrusion Cooking	Versatile, produces a variety of products, improves texture and bioavailability	High initial equipment cost, may degrade some heat-sensitive nutrients	At 118°C–122°C, screw speed 345–355 rpm, and feeder speed 5–15 kg/h	Sukumar and Athmaselvi (2019)
Puffing	Quick, improves texture and palatability, enhances sensory properties	High energy consumption, risk of nutrient degradation	At 230°C–250°C for very short time	Verma and Patel (2013)
Microwave Heating	Fast, energy-efficient, retains nutrients, uniform heating	Limited to small batches, uneven heating in some cases	720 W for 2–6 min	Ananthu *et al.* (2023)
Infrared Heating	Energy-efficient, reduces moisture effectively, retains nutrients	Requires specialized equipment	1 kW, peak infrared wavelength = 1.1–1.2 μm)	Kumari and Jayadeep (2024)
Baking	Versatile, produces value-added products like bread, cookies, and cakes	Loss of some heat-sensitive nutrients, requires precise temperature control	180°C-220°C for 15–50 min	Swami *et al.* (2013)
Boiling	Simple, improves digestibility, reduces anti-nutritional factors	Significant loss of water-soluble nutrients, time-consuming	At 100°C for 20–30 min	Okolo and Adejumo (2018)

11.3.1.2 Fluidized Bed Drying

Hot air is passed through the millet grains, creating a fluid-like state that allows for rapid and uniform drying. This method reduces time required for drying and energy consumption in comparison with traditional methods, which makes it a more efficient choice for large-scale processing (Setareh *et al.*, 2024).

11.3.1.3 Microwave Drying

Microwave drying is a more recent innovation in thermal processing. It uses microwave energy to heat water molecules within the grains, causing them to evaporate rapidly. Microwave drying offers various benefits, such as reduced drying times, decreased energy usage, and improved preservation of heat-sensitive nutrients like vitamins and antioxidants. Research has demonstrated that microwave drying can effectively preserve the bioactive compounds in millets, which are known for their health benefits (Radhika *et al.*, 2013).

11.3.2 Roasting

Roasting is another critical thermal process used to enhance the flavor, aroma, and texture of millets. The process involves heating the grains at high temperatures, which induces chemical processes like caramelization and maillard reaction (if reducing sugar is present). These reactions lead to development of desired color and flavor, making roasted millets more appealing for consumption and further processing.

11.3.2.1 Conventional Roasting

Traditional roasting methods involve heating millet grains in pans or drums, typically over an open flame or in a heated sand bed. This method has been used for centuries but often results in uneven roasting and potential nutrient loss due to uncontrolled heating. Conventional roasting is still practiced in many rural areas where modern equipment may not be available, but its limitations have prompted the development of more advanced roasting techniques (Shingare and Thorat, 2013).

11.3.2.2 Infrared Roasting

Infrared roasting is a contemporary method that utilizes infrared radiation to heat the millet grains. The infrared rays penetrate the grains, resulting in internal heating and ensuring a more consistent roasting. This method has numerous benefits compared to traditional roasting, such as shorter roasting times, improved control over the roasting process, and decreased nutrient loss. Research indicates that infrared roasting can greatly boost the antioxidant capacity of finger millets, further enhancing the health benefits (Shingare and Thorat, 2013).

11.3.2.3 Hot Air Roasting

In hot air roasting, millet grains are exposed to hot air in a controlled environment. This method allows for precise control over temperature and roasting time, resulting in uniformly roasted grains with consistent quality. Hot air roasting is a common method employed in the manufacturing of roasted millet flour, which serves as a healthy ingredient in a variety of food items like bread, cookies, and porridges (Chandrasekara and Shahidi, 2011).

11.3.3 Puffing

Puffing is a process in which grain is heated at a very high temperature for a very short period of time due to which the millet grains are expanded rapidly, creating a light and airy texture. Puffing not only enhances the texture of millets but also improves their digestibility by altering the starch structure, making them easier to digest and absorb.

11.3.3.1 Traditional Puffing

Traditional puffing involves heating millet grains in hot sand or an open flame until they expand and puff. While effective, this method often results in uneven puffing and a burnt flavor due to the lack of precise temperature control. Additionally, traditional puffing can lead to significant nutrient loss, particularly in heat-sensitive compounds (Mishra *et al.*, 2014).

11.3.3.2 Hot Air Puffing

Hot air puffing is a more controlled process in which millet grains are exposed to a stream of hot air, causing them to puff uniformly. This method reduces the risk of burning and nutrient loss while producing puffed grains with a consistent texture and flavor. Hot air puffing is a popular method used in making puffed millet snacks, which are gaining popularity as a healthier option to traditional snacks (Chandrasekara and Shahidi, 2011).

11.3.3.3 Microwave Puffing

Microwave puffing is a modern method that utilizes microwave energy to quickly heat and puff millet grains. This approach has numerous benefits, such as consistent puffing, reduced processing times, and minimal nutrient depletion. Microwave puffing has been shown to enhance the bioavailability of certain nutrients, such as proteins and polyphenols, making puffed millet products not only tasty but also nutritionally superior (Mishra *et al.*, 2014).

11.3.4 Extrusion

Extrusion is applied to create various products with diverse shapes, textures, and nutritional compositions. It is a very flexible type of thermal processing technique. In millet processing, extrusion is used to transform millet flour into products such as snacks, breakfast cereals, and pasta.

11.3.4.1 Single-Screw Extrusion

Single-screw extrusion is a process in which millet flour is fed into a barrel containing a rotating screw. The flour undergoes high temperature and pressure as it is pushed through the barrel, resulting in it cooking and expanding as it comes out of the die. This method is commonly used for producing puffed millet snacks and breakfast cereals. Single-screw extrusion is relatively simple and cost-effective, making it suitable for small to medium-scale production.

11.3.4.2 Twin-Screw Extrusion

Twin-screw extrusion is a more advanced process that uses two intermeshing screws to mix, cook, and shape the millet flour. This technique provides more precise control over the processing conditions, enabling the creation of a diverse array of products with different textures and shapes. Twin-screw extrusion is especially beneficial for decreasing anti-nutritional components like phytic acid and enhancing protein digestibility. It is commonly employed in the manufacturing of fortified millet products to combat nutritional deficiencies in at-risk populations.

11.3.4.3 High-Temperature Short-Time (HTST) Extrusion

HTST extrusion is a specific type of twin-screw extrusion that involves exposing the millet flour to very high temperatures for a short period. This method is particularly effective in producing products with desirable textures and shelf stability. HTST extrusion has been used to produce millet-based snack foods that are both crunchy and nutritious, appealing to health-conscious consumers (Nishani *et al.*, 2017).

11.4 Value Addition through Thermal Processing

Thermal processing technologies have played a significant role in the value addition of millets, transforming them from traditional staple grains into modern, value-added products. These technologies enhance the sensory properties of millets, making them more appealing to consumers, and improve their nutritional profile by enhancing the bioavailability of key nutrients.

11.4.1 Development of Millet-Based Snack Foods

The use of puffing and extrusion technologies is leading to development of various millet-based snack foods. These snacks are often marketed as healthy alternatives to conventional snacks, as they are rich in fiber, protein, and antioxidants. For example, puffed millet snacks are popular in many regions as they offer a crunchy texture and a light, airy feel that appeals to a wide range of consumers (Chandrasekara and Shahidi, 2011).

11.4.2 Fortified Millet Products

Thermal processing, particularly extrusion, has been used to develop fortified millet products that address specific nutritional deficiencies. For instance, extruded millet products fortified with vitamins and minerals are being developed to combat micronutrient deficiencies in vulnerable populations. These products are often used in school feeding programs and other nutritional interventions (Tamilselvan *et al.*, 2023).

11.4.3 Ready to Cook/Eat Products

Thermal processing has also helped in creating ready to cook/eat millet-based products to meet the increasing demand for convenient foods. For instance, roasted flour of millet can be used for making instant porridges or as an ingredient in bakery items, providing consumers with a fast and healthy meal choice (Rai *et al.*, 2023).

11.5 Challenges and Future Directions

While thermal processing technologies have significantly enhanced the utilization of millets, several challenges remain. One of the primary challenges is the potential loss of heat labile nutrients like flavor compounds and vitamins, during processing. The loss of nutrients can diminish the health benefits of millets, making it essential to optimize processing conditions to preserve these compounds. Future research should focus on developing processing techniques that minimize nutrient degradation while maximizing the benefits of thermal processing. For example, the combination of thermal processing techniques with other techniques like ultrasound processing or high-pressure processing, could help retain heat-sensitive nutrients while still achieving the desired product quality (Zhang *et al.*, 2006). Achieving consistent product quality is another challenge in millet processing. Variations in millet grain composition, processing conditions, and equipment can lead to inconsistencies in the final product, affecting its marketability. Standardizing processing protocols and improving equipment design could help address these issues and ensure consistent product quality (Mujumdar, 2006). Sustainability is becoming an increasingly important consideration in food processing. Thermal processing technologies, particularly those that require high energy inputs, can have a significant environmental impact. Developing energy-efficient processing techniques and exploring alternative energy sources, such as solar energy for drying, could help reduce the environmental footprint of millet processing (Geetanjali and Singh, 2024).

11.6 Conclusion

Millets are the crops which contains high amount of nutrients and are adaptable to various climates and requires minimal irrigation. However, its flour has a short shelf life due to high lipid and enzymatic activity, limiting its widespread use. Various processing methods are developed for enhancing the nutritional and functional qualities of finger millet while reducing nutrient loss and energy consumption. Thermal processing technologies have revolutionized the way millets are processed and utilized, transforming them from traditional staples into various ready to cook/eat food products. From drying and roasting to puffing and extrusion, each thermal processing technique offers unique advantages in improving the quality, shelf life, and marketability of millets. As consumer demand for healthy, sustainable, and convenient food options continues to grow, the role of thermal processing in millet utilization will become increasingly important. Further research and innovation in this field are crucial for fully realizing the potential of millets as a valuable food source, thereby enhancing global food security and nutrition. Additional large-scale studies are needed to make processing techniques economically feasible, with a focus on the processing and utilization of finger millets. It is also important to optimize processing conditions for stabilizing finger millet flour.

References

Ananthu, R., Buvaneswaran, M., Meena, L., & Sunil, C. K. (2023). Microwave treatment effect on functional and pasting properties, and storage stability of white finger millet. *Journal of Food Process Engineering*, 46(7), e14341. https://doi.org/10.1111/jfpe.14341

Angioloni, A., & Collar, C. (2012). Effects of pressure treatment of hydrated oat, finger millet and sorghum flours on the quality and nutritional properties of composite wheat breads. *Journal of Cereal Science*, 56(3), 713–719. https://doi.org/10.1016/j.jcs.2012.08.001

Arsha, S., Abdullah, S., Navya, K. P., Sunooj, K. V., Abhishek, S., & Sudheer, K. P. (2025). Cold plasma-induced modification of finger millet (*Eleusine coracana*) flour: Unveiling potential for anti-nutrient reduction and functional enrichment. *European Food Research and Technology*. https://doi.org/10.1007/s00217-025-04689-y

Awad, T. S., Moharram, H. A., Shaltout, O. E., Asker, D., & Youssef, M. M. (2012). Applications of ultrasound in analysis, processing and quality control of food: A review. *Food Research International*, 48(2), 410–427.

Barba, F., Sant'Ana, A., Orlien, V., & Koubaa, M. (2017). *Innovative Technologies for Food Preservation: Inactivation of Spoilage and Pathogenic Microorganisms* (pp. 133–139). Academic Press.

Bisht, B., Bhatnagar, P., Gururani, P., Kumar, V., Tomar, M. S., Sinhmar, R., Rathi, N., & Kumar, S. (2021). Food irradiation: Effect of ionizing and non-ionizing radiations on preservation of fruits and vegetables – A review. *Trends in Food Science and Technology*, 114, 372–385.

Chandra, D., Chandra, S., Pallavi, & Sharma, A. K. (2016). Review of finger millet (*Eleusine coracana* (L.) Gaertn): A powerhouse of health benefiting nutrients. *Food Science and Human Wellness*, 5(3):149–155. https://doi.org/10.1016/j.fshw.2016.05.004.

Chandrasekara, A., & Shahidi, F. (2011). Determination of antioxidant activity in free and hydrolyzed fractions of millet grains and characterization of their phenolic profiles by HPLC-DAD-ESI-MSn. *Journal of Functional Foods*, 3(3), 144–158.

Cormick, G., & Belizán, J. M. (2019). Calcium intake and health. *Nutrients*, 11(7), 1606.

Cui, R., & Zhu, F. (2020). Effect of ultrasound on structural and physicochemical properties of sweet potato and wheat flours. *Ultrasonics Sonochemistry*, 66, 105–118.

Diouf, A., Sarr, F., Sene, B., Ndiaye, C., Fall, S. M., & Ayessou, N. C. (2019). Pathways for reducing anti-nutritional factors: Prospects for *Vigna unguiculata*. *Journal of Nutrition Health and Food Science*, 7(2), 1–10.

Geetanjali, & Singh, R. (2024). Sustainability and environmental issues in food processing. In *Nanobiotechnology for Food Processing and Packaging* (pp. 209–222). Academic Press.

Gomez, M., & Martinez, M. M. (2016). Changing four functionalities through physical treatments for the production of gluten-free baking goods. *Journal of Cereal Science*, 67, 68–74. https://doi.org/10.1016/j.jcs.2015.07.009.

Hayes, F., Harmens, H., Sharps, K., & Radbourne, A. (2020). Ozone dose-response relationships for tropical crops reveal potential threat to legume and wheat production, but not to millets. *Scientific African*, 9, e00482.

Hwang, H. J., Cheigh, C. I., & Chung, M. S. (2018). Comparison of bactericidal effects of two types of pilot-scale intense-pulsed-light devices on cassia seeds and glutinous millet. *Innovative Food Science & Emerging Technologies*, 49, 170–175.

Jeon, M. S., Park, K. M., Yu, H., Park, J. Y., & Chang, P. S. (2019). Effect of intense pulsed light on the deactivation of lipase: Enzyme-deactivation kinetics and tertiary structural changes by fragmentation. *Enzyme and Microbial Technology*, 124, 63–69. https://doi.org/10.1016/j.enzmictec.2019.02.001.

Joshi, T. J., Singh, S. M., & Rao, P. S. (2023). Novel thermal and non-thermal millet processing technologies: Advances and research trends. *European Food Research and Technology*, 249, 1149–1160. https://doi.org/10.1007/s00217-023-04227-8.

Kabeer, S., Govindarajan, N., Radhakrishnan, P., Essa, M. M., & Qoronfleh, M. W. (2023). Effect of drying technique on physiochemical and nutritional properties of *Eleusine coracana* (finger millet) porridge powder. *Journal of Food Science and Technology*, 60(12), 3024–3034.
Kramer, B., & Muranyi, P. (2013). Effect of pulsed light on structural and physiological properties of *Listeria innocua* and *Escherichia coli*. *Journal of Applied Microbiology*, 116(3), 596–611. https://doi.org/10.1111/jam.12394.
Kumari, V., & Jayadeep, A. (2024). Effect of infrared heating on finger millet in unhydrated and hydrated conditions on physical, functional, pasting, and phenolic profile. *Journal of Cereal Science*, 116, 103834.
Lokeswari, R., Sharanyakanth, P. S., Jaspin, S., & Mahendran, R. (2021). Cold-plasma effects on changes in physical, nutritional, hydration, and pasting properties of pearl millet (*Pennisetum Glaucum*). *IEEE Transactions on Plasma Science*, 49(5), 1745–1751.
Mahendran, R., Ramanan, K. R., Barba, F. J., Lorenzo, J. M., Fernandez, O. L., Munekata, P. E. S., Roohinejad, S., Sant'Ana, A. S., & Tiwari, B. K. (2019). Recent advances in the application of pulsed light processing for improving food safety and increasing shelf life. *Trends in Food Science and Technology*, 88, 67–79. https://doi.org/10.1016/j.tifs.2019.03.010.
Mahmoud, N. S., Awad, S. H., Madani, R. M. A., Osman, F. A., Elmamoun, K., & Hassan, A. B. (2015). Effect of γ radiation processing on fungal growth and quality characteristics of millet grains. *Food Science and Nutrition*, 4(3), 342–347. https://doi.org/10.1002/fsn3.295.
Mishra, G., Joshi, D. C., & Panda, B. K. (2014). Popping and puffing of cereal grains: A review. *Journal of Grain Processing and Storage*, 1(2), 34–46.
Mujumdar, A. S. (2006). *Handbook of Industrial Drying*. CRC Press.
Mollakhalili-Meybodi, N., Nejati, R., Sayadi, M., & Nematollahi, A. (2022). Novel non-thermal food processing practices: Their influences on nutritional and technological characteristics of cereal proteins. *Food Science & Nutrition*, 10(6), 1725–1744.
Nazari, B., Mohammadifar, M. A., Shojaee-Aliabadi, S., Feizollahi, E., & Mirmoghtadaie, L. (2018). Effect of ultrasound treatments on functional properties and structure of millet protein concentrate. *Ultrasonics Sonochemistry*, 41, 382–388. https://doi.org/10.1016/j.ultsonch.2017.10.002
Nishani, S., Rudra, S. G., & Verghese, E. (2017). Effect of extrusion processing variables on finger millet flours with respect to their functional properties using response surface methodology. *Current Journal of Applied Science and Technology*, 23(2), 1–9. https://doi.org/10.9734/CJAST/2017/34495
Okolo, J. C., & Adejumo, B. A. (2018). Functional qualities of thermally processed finger millet (*Eleusine Coracana* L. Gaertn) flour. *International Journal of Research in Engineering and Science*, 6(5), 22–26.
Oluwole, O., Fernando, W. M. B., Audain, K., Fasanmade, O., Ijabadeniyi, O., Falade, K., Yaakwaah, A. M., & Jayasena, V. (2019). The effects of germination and roasting on nutraceutical and antioxidant properties of *Jirani* variety of millet. *American Journal of Food Science and Technology*, 7(6), 234–241.
Premi, M., & Khan, K. A. (2018). Irradiation Technology: Processing of Fruits and Vegetables. In *Technological Interventions in the Processing of Fruits and Vegetables* (pp. 209–226). Apple Academic Press.
Radhika, G. B., Satyanarayana, S. V., & Rao, D. G. (2013). An investigation on drying of millets in a microwave oven. *International Journal of Emerging Technology and Advanced Engineering*, 3, 583–589.

Rai, S., Ramachandran, P., & Sangeeta, S. (2023). Value Addition of Cereal and Millet Processing Industrial Waste. In *Wealth Out of Food Processing Waste* (pp. 41–72). CRC Press.

Ramakrishnan, S. R., Antony, U., & Kim, S. J. (2023). Non-thermal process technologies: Influences on nutritional and storage characteristics of millets. *Journal of Food Process Engineering*, 46(10), e14215.

Rani, S., Singh, R., Sehrawat, R., Kaur, B. P., & Upadhyay, A. (2018). Pearl Millet processing: A review. *Nutrition & Food Science*, 48(1), 30–44. https://doi.org/10.1108/NFS-04-2017-0070

Rao, B., Dayakar, Bhaskarachary, K., Christina, A., Sudha, G. D., Vilas, D. G., & Tonapi, A. (2017). *Nutritional and Health Benefits of Millets* (p. 112). ICAR_Indian Institute of Millets Research (IIMR).

Reddy, C. K., & Viswanath, K. K. (2019). Impact of γ-irradiation on physicochemical characteristics, lipoxygenase activity and antioxidant properties of finger millet. *Journal of Food Science and Technology*, 56(5), 2651–2659. https://doi.org/10.1007/s13197-019-03753-2

Sarangapani, C., Thirumdas, R., Devi, Y., Trimukhe, A., Deshmukh, R. R., & Annapure, U. S. (2016). Effect of low-pressure plasma on physico–chemical and functional properties of parboiled rice flour. *LWT - Food Science and Technology*, 69, 482–489.

Setareh, M., Assari, M. R., Tabrizi, H. B., & Zadeh, A. M. (2024). Experimental and drying kinetics study on millet particles by a pulsating fluidized bed dryer. *Heliyon*, 10(13), e33680.

Shingare, S. P., & Thorat, B. N. (2013). Effect of drying temperature and pretreatment on protein content and color changes during fluidized bed drying of finger millets (ragi, *Eleusine coracana*) sprouts. *Drying Technology*, 31(5), 507–518.

Sirohi, R., Tarafdar, A., Gaur, V. K., Singh, S., Sindhu, R., Rajasekharan, R., Madhavan, A., Binod, P., Kumar, S., & Pandey, A. (2021). Technologies for disinfection of food grains: Advances and way forward. *Food Research International*, 145, 110396. https://doi.org/10.1016/j.foodres.2021.110396.

Somarajan, A., & Morya, S. (2022). A review on finger millet properties, processing, health benefits, and applications. *The Pharma Innovation*, 11(7), 3388–3395. https://doi.org/10.22271/tpi.2022.v11.i7aq.14538.

Sruthi, N. U., & Rao, P. S. (2021). Effect of processing on storage stability of millet four: A review. *Trends Food Science and Technology*, 112, 58–74. https://doi.org/10.1016/j.tifs.2021.03.043.

Sukumar, A., & Athmaselvi, K. A. (2019). Optimization of process parameters for the development of finger millet based multigrain extruded snack food fortified with banana powder using RSM. *Journal of Food Science and Technology*, 56(2), 705–712. https://doi.org/10.1007/s13197-018-3527-x

Sullivan, J. O., Murray, B., Flynn, C., & Norton, I. (2016). The effect of ultrasound treatment on the structural, physical and emulsifying properties of animal and vegetable proteins. *Food Hydrocolloids*, 53, 141–154. https://doi.org/10.1016/j.foodhyd.2015.02.009.

Swami, S. B., Thakor, N. J., & Shinde, U. N. (2013). Effect of baking temperature and oil content on quality of finger millet flour cookies. *Journal of Agricultural Engineering*, 50(4), 32–38.

Tamilselvan, T., Sharma, S., & Prabhasankar, P. (2023). Finger Millet (Eleusine coracana): Phytochemical Profile, Potential Health Benefits, and Techno-Functional Properties. In *Nutri-Cereals* (pp. 119–139). CRC Press.

Taylor, J. R. N., & Kruger, J. (2016). Millets. *Encyclopedia of Food and Health*, 748–757. https://doi.org/10.1016/b978-0-12-384947-2.00466-9
Vanga, S. K., Wang, J., Jayaram, S., & Raghavan, V. (2021). Effects of pulsed electric fields and ultrasound processing on proteins and enzymes: A review. *Processes*, 9(4), 722.
Verma, V., & Patel, S. (2013). Value added products from nutri-cereals: Finger millet (*Eleusine coracana*). *Emirates Journal of Food and Agriculture*, 25(3), 169–176.
Xue, W., Macleod, J., & Blaxland, J. (2023). The use of ozone technology to control microorganism growth, enhance food safety and extend shelf life: A promising food decontamination technology. *Foods*, 12(4), 814. https://doi.org/10.3390/foods12040814.
Yadav, S., Mishra, S., & Pradhan, R. C. (2021). Ultrasound-assisted hydration of finger millet (*Eleusine coracana*) and its effects on starch isolates and antinutrients. *Ultrasonics Sonochemistry*, 73, 105542. https://doi.org/10.1016/j.ultsonch.2021.105542.
Zhang, M., Tang, J., Mujumdar, A. S., & Wang, S. (2006). Trends in microwave-related drying of fruits and vegetables. *Trends in Food Science & Technology*, 17(10), 524–534.
Zhang, Y., Jing, X., Chen, Z., & Wang, X. (2023). Effects of moderate-intensity pulsed electric field on the structure and physicochemical properties of foxtail millet (*Setaria italica*) prolamin. *Cereal Chemistry*, 100(2), 360–370.
Zhu, F., & Li, H. (2019). Modification of quinoa four functionality using ultrasound. *Ultrasonics Sonochemistry*, 52, 305–310. https://doi.org/10.1016/j.ultsonch.2018.11.027.

Index

Note: **Bold** and *italic* page numbers refer to tables and figures, respectively.

O

P

Y

Z

For Product Safety Concerns and Information please contact our EU
representative GPSR@taylorandfrancis.com
Taylor & Francis Verlag GmbH, Kaufingerstraße 24, 80331 München, Germany

www.ingramcontent.com/pod-product-compliance
Ingram Content Group UK Ltd.
Pitfield, Milton Keynes, MK11 3LW, UK
UKHW022210190926
475702UK00003B/292
* 9 7 8 1 0 3 2 8 7 8 7 4 4 *